Health through Will Power

By

James J. Walsh

HEALTH THROUGH WILL POWER

CHAPTER I

THE WILL IN LIFE

"What he will he does and does so much
That proof is called impossibility."
Troilus and Cressida.

The place of the will in its influence upon health and vitality has long been recognized, not only by psychologists and those who pay special attention to problems of mental healing, but also, as a rule, by physicians and even by the general public. It is, for instance, a well-established practice, when two older folk, near relatives, are ill at the same time, or even when two younger persons are injured together and one of them dies, or perhaps has a serious turn for the worse, carefully to keep all knowledge of it from the other one. The reason is a very definite conviction that in the revulsion of feeling caused by learning of the fatality, or as a result of the solicitude consequent upon hearing that there has been a turn for the worse, the other patient's chances for recovery would probably be seriously impaired. The will to get better, even to live on, is weakened, with grave consequences. This is no mere popular impression due to an exaggeration of sympathetic feeling for the patient. It has been noted over and over again, so often that it evidently represents some rule of life, that whenever by inadvertence the serious condition or death of the other was made known, there was an immediate unfavorable development in the case which sometimes ended fatally, though all had been going well up to that time. This was due not merely to the shock, but largely to the "giving up", as it is called, which left the surviving patient without that stimulus from the will to get well which means so much.

It is surprising to what an extent the will may affect the body, even under circumstances where it would seem impossible that physical factors could any longer have any serious influence. We often hear it said that certain people are "living on their wills", and when they are of the kind who take comparatively little food and yet succeed in accomplishing a great deal of work, the truth of the expression comes home to us rather strikingly. The expression is usually considered, however, to be scarcely more than a formula of words elaborated in order to explain certain of these exceptional cases that seem to need some special explanation. The possibility of the human will of itself actually prolonging existence beyond the time when, according to all reason founded on physical grounds, life should end, would seem to most people to be quite out of the question. And yet there are a number of striking cases on record in which the only explanation of the continuance of life would seem to be that the will to live has been so strongly aroused that life was prolonged beyond even expert expectation. That the will was the survival factor in the case is clear from the fact that as soon as this active willing process ceased, because the reason that had aroused it no longer existed, the individuals in question proceeded to reach the end of life rapidly from the physical factors already at work and which seemed to portend inevitably an earlier dissolution than that which happened. Probably a great many physicians know of striking examples of patients who have lived beyond the time when ordinarily death would be expected, because they were awaiting the arrival of a friend from a distance who was known to be coming and whom the patient wanted very much to see. Dying mothers have lived on to get a last embrace of a son or daughter, and wives have survived to see their husbands for a last parting though it seemed impossible that they should do so, so far as their physical condition was concerned and then expired within a short time. Of course there are any number of examples in which this has not been true, but then that is only a proof of the fact that the great majority of mankind do not use their wills, or perhaps, having appealed to them for help during life never or but slightly, are not prepared to make a definite serious call on them toward the end. I am quite sure, however, that a great many country physicians particularly can tell stories of incidents that to them were proofs that the will can resist even the approach of death for some time, though just as soon as the patients give up, death comes to them.

Professor Stokes, the great Irish clinician of the nineteenth century, to whom we owe so much of our knowledge of the diseases of the heart and lungs, and whose name is enshrined in terms commonly used in medicine in connection with these diseases, has told a striking story of his experiences in a Dublin hospital that illustrates this very well. An old Irishman, who had been a soldier in his younger years and had been wounded many times, was in the hospital ill and manifestly dying. Professor Stokes, after a careful investigation of his condition, declared that he could not live a week, though at the end of that time the old soldier was still hanging on to life, ever visibly sinking. Stokes assured the students who were making the rounds of his wards with him that the old man had at most a day or two more to live, and yet at the end of some days he was still there to greet them on their morning visits. After the way of medical students the world over, though without any of that hard-heartedness that would be supposed ordinarily to go with such a procedure, for they were interested in the case as a medical problem, the students began to bet how long the old man would live.

Finally, one day the old man said to Stokes in his broadest brogue: "Docther, you must keep me alive until the first of the month, because me pension for the quarther is then due, and unless the folks have it, shure they won't have anything to bury me with."

The first of the month was some ten days away. Stokes said to his students, though of course not in the hearing of the patient, that there was not a chance in the world, considering the old soldier's physical condition, that he would live until the first of the month. Every morning, in spite of this, when they came in, the old man was still alive and there was even no sign of the curtains being drawn around his bed as if the end were approaching. Finally on the morning of the first of the month, when Stokes came in, the old pensioner said to him feebly, "Docther, the papers are there. Sign them! Then they'll get the pension. I am glad you kept me alive, for now they'll surely have the money to bury me." And then the old man, having seen the signature affixed, composed himself for death and was dead in the course of a few hours. He had kept himself alive on his will because he had a purpose in it, and once that purpose was fulfilled, death was welcome and it came without any further delay.

There is a story which comes to us from one of the French prisons about the middle of the nineteenth century which illustrates forcibly the same power of the will to maintain life after it seemed sure, beyond peradventure, that death must come. It was the custom to bury in quicklime in the prison yard the bodies of all the prisoners who died while in custody. The custom still survives, or did but twenty years ago, even in English prisons, for those who were executed, as readers of Oscar Wilde's "Ballad of Reading Gaol" will recall. Irish prisons still keep up the barbarism, and one of the reasons for the bitterness of the Irish after the insurrection of in Dublin was the burial of the executed in quicklime in the prison yard. The Celtic mind particularly revolts at the idea, and it happened that one of the prisoners in a certain French prison, a Breton, a Celt of the Celts, was deeply affected by the thought that something like this might happen to him. He was suffering from tuberculosis at a time when very little attention was paid to such ailments in prisoners, for the sooner the end came, the less bother there was with them; but he was horrified at the thought that if he died in prison his body would disappear in the merciless fire of the quicklime.

So far as the prison physician could foresee, the course of his disease would inevitably reach its fatal termination long before the end of his sentence. In spite of its advance. however, the prisoner himself declared that he would never permit himself to die in prison and have his body face such a fate. His declaration was dismissed by the physician with a shrug and the feeling that after all it would not make very much difference to the man, since he would not be there to see or feel it. When, however, he continued to live, manifestly in the last throes of consumption, for weeks and even months after death seemed inevitable, some attention was paid to his declaration in the matter and the doctors began to give special attention to his case. He lived for many months after the time when, according to all ordinary medical knowledge, it would seem he must surely have died. He actually outlived the end of his sentence, had arrangements made to move him to a house just beyond the prison gate as soon as his sentence had expired, and according to the story, was dead within twenty-four hours after the time he got out of prison and thus assured his Breton soul of the fact that his body would be given, like that of any Christian, to the bosom of mother earth.

But there are other and even more important phases of the prolongation of life by the will that still better illustrates its power. It has often been noted that men who have had extremely busy lives, working long hours every day, often sleeping only a few hours at night, turning from one thing to another and accomplishing so much that it seemed almost impossible that one man could do all they did, have lived very long lives. Men like Alexander Humboldt, for instance, distinguished in science in his younger life, a traveler for many years in that hell-hole of the tropics, the region around Panama and Central America, a great writer whose books deeply influenced his generation in middle age. Prime Minister of Prussia as an older man, lived to be past ninety, though he once confessed that in his forties he often slept but two or three hours a night and sometimes took even that little rest on a sofa instead of a bed. Leo XIII at the end of the nineteenth century was just such another man. Frail of body, elected Pope at sixty-four, it was thought that there would soon be occasion for another election; he did an immense amount of work, assumed successfully the heaviest responsibilities, and outlived the years of Peter in the Papal chair, breaking all the prophecies in that regard and not dying till he was ninety-three.

Many other examples might be cited. Gladstone, always at work, probably the greatest statesman of the nineteenth century in the better sense of that word, was a scholar almost without a peer in the breadth of his scholarly attainments, a most interesting writer on multitudinous topics, keen of interest for everything human and always active, and yet he lived well on into the eighties. Bismarck and Von Moltke, who assumed heavier responsibilities than almost any other men of the nineteenth century, saw their fourscore years pass over them a good while before the end came. Bismarck remarked on his eighty-first birthday that he used to think all the good things of life were confined to the first eighty years of life, but now he knew that there were a great many good things reserved for the second eighty years. I shall never forget sitting beside Thomas Dunn English, the American poet, at a banquet of the Alumni of the University of Pennsylvania, when he repeated this expression for us; at the time he himself was well past eighty. He too had been a busy man and yet rejoiced to be with the younger alumni at the dinner.

My dear old teacher, Virchow, of whom they said when he died that four men died, for he was distinguished not only as a pathologist, which was the great life-work for which he was known, but as an anthropologist, a historian of medicine and a sanitarian, was at seventy-five actively accomplishing the work of two or three men. He died at eighty-one, as the result of a trolley injury, or I could easily imagine him alive even yet. Von Ranke, the great historian of the popes, began a universal history at the age of ninety which was planned to be complete in twelve volumes, one volume a year to be issued. I believe that he lived to finish half a dozen of them. I have some dear friends among the medical profession in America who are in their eighties and nineties, and all of them were extremely busy men in their middle years and always lived intensely active lives. Stephen Smith and Thomas Addis Emmet, John W. Gouley, William Hanna Thompson, not long dead, and S. Weir Mitchell, who lived to be past eighty-five, are typical examples of extremely busy men, yet of extremely long lives.

All of these men had the will power to keep themselves busily at work, and their exercise of that will power, far from wearing them out, actually seemed to enable them to tap reservoirs of energy that might have remained latent in them. The very intensiveness of their will to do seemed to exert an extensive influence over their lives, and so they not only accomplished more but actually lived longer. Hard work, far from exhausting, has just the opposite effect. We often hear of hard work killing people, but as a physician I have carefully looked into a number of these cases and have never found one which satisfied me as representing exhaustion due to hard work. Insidious kidney disease, rheumatic heart disease, the infections of which pneumonia is a typical example, all these have been the causes of death and not hard work, and they may come to any of us. They are just as much accidents as any other of the mischances of life, for it is as dangerous to be run into by a microbe as by a trolley car. Using the will in life to do all the work possible only gives life and gives it more abundantly, and people may rust out, that is be hurt by rest, much sooner than they will wear out.

Here, then, is a wellspring of vitality that checks for a time at least even lethal changes in the body and undoubtedly is one of the most important factors for the prolongation of life. It represents the greatest force for health and power of accomplishment that we have. Unfortunately, in recent years, it has been neglected to a great extent for a number of reasons. One of these has been the discussions as to the freedom of the will and the very common teaching of determinism which seemed to eliminate the will as an independent faculty in life. While this affected only the educated classes who had received the higher education, their example undoubtedly was pervasive and influenced a great many other people. Besides, newspaper and magazine writers emphasized for popular dissemination the ideas as to absence of the freedom of the will which created at least an unfortunate attitude of mind as regards the use of the will at its best and tended to produce the feeling that we are the creatures of circumstances rather than the makers of our destiny in any way, or above all, the rulers of ourselves, including even to a great extent our bodily energies.

Even more significant than this intellectual factor, in sapping will power has been the comfortable living of the modern time with its tendency to eliminate from life everything that required any exercise of the will. The progress which our generation is so prone to boast of concerns mainly this making of people more comfortable than they were before. The luxuries of life of a few centuries ago have now become practically the necessities of life of to-day. We are not asked to stand cold to any extent, we do not have to tire ourselves walking, and bodily labor is reserved for a certain number of people whom we apparently think of as scarcely counting in the scheme of humanity. Making ourselves comfortable has included particularly the removal of nearly all necessity for special exertion, and therefore of any serious exercise of the will. We have saved ourselves the necessity for expending energy, apparently with the idea that thus it would accumulate and be available for higher and better purposes.

The curious thing with regard to animal energy, however, is that it does not accumulate in the body beyond a certain limited extent, and all energy that is manufactured beyond that seems to have a definite tendency to dissipate itself throughout the body, producing discomfort of various kinds instead of doing useful work. The process is very like what is called short-circuiting in electrical machinery, and this enables us to understand how much harm may be done. Making ourselves comfortable, therefore, may in the end have just exactly the opposite effect, and often does. This is not noted at first, and may escape notice entirely unless there is an analysis of the mode of life which is directed particularly to finding out the amount of exertion of will and energy that there is in the daily round of existence.

The will, like so many other faculties of the human organism, grows in power not by resting but by use and exercise. There have been very few calls for serious exercise of the will left in modern life and so it is no wonder that it has dwindled in power. As a consequence, a good deal of the significance of the will in life has been lost sight of. This is unfortunate, for the will can enable us to tap sources of energy that might otherwise remain concealed from us. Professor William James particularly called attention to the fact, in his well-known essay on "The Energies of Men", that very few people live up to their maximum of accomplishment or their optimum of conduct, and that indeed "as a rule men habitually use only a small part of the powers which they actually possess and which they might use under appropriate conditions."

It is with the idea of pointing out how much the will can accomplish in changing things for the better that this volume is written. Professor James quoted with approval Prince Pueckler-Muskau's expression, "I find something very satisfactory in the thought that man has the power of framing such props and weapons out of the most trivial materials, indeed out of nothing, merely by the force of his will, which thereby truly deserves the name of omnipotent." [Footnote]

It is this power, thus daringly called omnipotent, that men have not been using to the best advantage to maintain health and even to help in the cure of disease, which needs to be recalled emphatically to attention. The war has shown us in the persons of our young soldiers that the human will has not lost a single bit of its pristine power to enable men to accomplish what might almost have seemed impossible. One of the heritages from the war should be the continuance of that fine use of the will which military discipline and war's demands so well brought into play. Men can do and stand ever so much more than they realize, and in the very doing and standing find a satisfaction that surpasses all the softer pleasant feelings that come from mere comfort and lack of necessity for physical and psychical exertion. Their exercise of tolerance and their strenuous exertion, instead of exhausting, only makes them more capable and adds to instead of detracting from their powers.

How much this discipline and training of the will meant for our young American soldiers, some of whom were raised in the lap of luxury and almost without the necessity of ever having had to do or stand hard things previously in their lives, is very well illustrated by a letter quoted by Miss Agnes Repplier in the Century for December. It is by no means unique or even exceptional. There were literally thousands of such letters written by young officers similarly circumstanced, and it is only because it is typical and characteristic of the spirit of all of these young men that I quote it here. Miss Repplier says that it came from "a young American lieutenant for whom the world had been from infancy a perilously pleasant place." He wrote home in the early spring of :

"It has rained and rained and rained. I am as much at home in a mud puddle as any frog in France, and I have clean forgotten what a dry bed is like. But I am as fit as a fiddle and as hard as nails. I can eat scrap iron and sleep standing. Aren't there things called umbrellas, which you pampered civilians carry about in showers?" If we can secure the continuance of this will exertion, life will be ever so much heartier and healthier and happier than it was before, so that the war shall have its compensations.

CHAPTER II

DREADS

"O, know he is the bridle of your will.
There's none but asses will be bridled so."
A Comedy of Errors.

It must be a surprise to most people, after the demonstration of the power of the will in the preceding chapter, that so many fail to make use of it. Indeed, the majority of mankind are quite unable to realize the store of energy for their health and strength and well-being which is thus readily available, though so often unused or called upon but feebly. The reason why the will is not used more is comparatively easy to understand, however, once its activity in ordinary conditions of humanity is analyzed a little more carefully. The will is unfortunately seldom permitted to act freely. Brakes are put on its energies by mental states of doubt and hesitation, by contrary suggestion, and above all by the dreads which humanity has allowed to fasten themselves on us until now a great many activities are hampered. There is the feeling that many things cannot be done, or may be accomplished only at the cost of so much effort and even hardship that it would be hopeless for any but those who are gifted with extremely strong wills to attempt them. People grow afraid to commit themselves to any purpose lest they should not be able to carry it out. Many feel that they would never be able to stand what others have stood without flinching and are persuaded that if ever they were placed in the position where they had to withstand some of the trials that they have heard of they would inevitably break down under the strain.

Just as soon as a human being loses confidence that he or she may be able to accomplish a certain thing, that of itself is enough to make the will ever so much less active than it would otherwise be. It is like breaking a piece of strong string: those who know how wrap it around their fingers, then jerk confidently and the string is broken. Those who fear that they may not be able to break it hesitate lest they should hurt themselves and give a half-hearted twitch which does not break the string; the only thing they succeed in doing is in hurting themselves ever so much more than does the person who really breaks it. After that abortive effort, they feel that they must be different from the others whose fingers were strong enough to break the string, and they hesitate about it and will probably refuse to make the attempt again.

It is a very old story, this of dreads hampering the activities of mankind with lack of confidence, and the fear of failure keeping people from doing things. One of his disciples, according to a very old tradition, once asked St. Anthony the Hermit what had been the hardest obstacle that he found on the road to sanctity. The story has all the more meaning for us here if we recall that health and holiness are in etymology the same. St. Anthony, whose temptations have made him famous, was over a hundred at the time and had spent some seventy years in the desert, almost always alone, and probably knew as much about the inner workings of human nature from the opportunities for introspection which he had thus enjoyed as any human being who ever lived. His young disciple, like all young disciples, wanted a short cut on the pathway that they were both traveling. The old man said to him, "Well, I am an old man and I have had many troubles, but most of them never happened."

Many a nightmare of doubt and hesitation disappears at once if the dread of it is overcome. The troubles that never happen, if dwelt upon, paralyze the will until health and holiness become extremely difficult of attainment.

There is the secret of the failure of a great many people in life in a great many ways. They fear the worst, dread failure, dampen their own confidence, and therefore fritter away their own energy. Anything that will enable them to get rid of the dreads of life will add greatly to their power to accomplish things inside as well as outside their bodies. Well begun is half done, and tackling a thing confidently means almost surely that it will be accomplished. If the dread of failure, the dread of possible pain in its performance, the dread of what may happen as a result of activity, if all these or any of them are allowed to obtrude themselves, then energy is greatly lessened, the power to do things hampered and success becomes almost impossible. This is as true in matters of health and strength as it is with regard to various external accomplishments. It takes a great deal of experience for mankind to learn the lesson that their dreads are often without reality, and some men never learn it.

Usually when the word dreads is used, it is meant to signify a series of psychic or psychoneurotic conditions from which sensitive, nervous people suffer a great deal. There is, for instance, the dread of dirt called learnedly misophobia, that exaggerated fear that dirt may cling to the hands and prove in some way deleterious which sends its victims to wash their hands from twenty to forty times a day. Not infrequently they wash the skin pretty well off or at least produce annoying skin irritation as the result of their feeling. There are many other dreads of this kind. Some of them seem ever so much more absurd even than this dread of dirt. Most of us have a dread of heights, that is, we cannot stand on the edge of a height and look down without trembling and having such uncomfortable feelings that it is impossible for us to stay there any length of time. Some people also are unable to sit in the front row of the balcony of a theater or even to kneel in the front row of a gallery at church without having the same dread of heights that comes to others at the edge of a high precipice. I have among my patients some clergymen who find it extremely difficult to stand up on a high altar, though, almost needless to say, the whole height is at most five or six ordinary steps.

Then there are people who have an exaggerated dread of the dark, so that it is quite impossible for them to sleep without a light or to sleep alone. Sometimes such a dread is the result of some terrifying incident, as the case in my notes in which the treasurer of a university developed an intense dread of the dark which made sleep impossible without a light, after he had been shot at by a burglar who came into his room and who answered his demand, "Who is that?" by a bullet which passed through the head of the bed. Most of the skotophobists, the technical name for dark-dreaders, have no such excuse as this one. Victims of nervous dreads have as a rule developed their dread by permitting some natural feeling of minor importance to grow to such an extent that it makes them very miserable.

Some cannot abide a shut-in place. Philip Gilbert Hamerton, the English writer and painter, often found a railroad compartment in the English cars an impossible situation and had to break his journey in order to get over the growing feeling of claustrophobia, the dread of shut-in places, which would steal over him.

There are any number of these dreads and, almost needless to say, all of them may interfere with health and the pursuit of happiness. I have seen men and women thrown into a severe nervous state with chilly feelings and cold sweat as the result of trying to overcome one of these dreads. They make it impossible for their victims to do a great many things that other people do readily, and sadly hamper their wills. There is only one way to overcome these dreads, and that is by a series of acts in the contrary direction until a habit of self-control with regard to these haunting ideas is secured. All mankind, almost without exception, has a dread of heights, and yet many thousands of men have in recent years learned to work on high buildings without very much inconvenience from the dread. The wages are good, they want to work this way, and the result is they take themselves in hand and gradually acquire self-control. I have had many of them tell me that at first they were sure they would never be able to do it, but the gradual ascent of the building as the work proceeded accustomed them to height, and after a while it became almost as natural to work high up in the air as on the first or second story of a building or even on the level ground.

The overcoming of these dreads is not easy unless some good reason releases the will and sets it to exerting its full power. When this is the case, however, the dread is overcome and the brake lifted after some persistence, with absolute assurance. Men who became brave soldiers have been known to have had a great dread of blood in early life. Some of our best surgeons have had to leave the first operation that they ever saw or they would have fainted, and yet after repeated effort they have succeeded in overcoming this sensitiveness. As a matter of fact, most people suffer so much from dreads because they yielded to a minor dread and allowed a bad habit to be formed. It is a question of breaking a bad habit by contrary acts rather than of overcoming a natural disposition. Many of those who are victims have the feeling that they cannot be expected to conquer nature this way. As a result, they are so discouraged at the very idea that success is dubious and practically impossible from their very attitude of mind; but it is only the second nature of a habit that they have to overcome, and this is quite another matter, for exactly contrary acts to these which formed a habit will break it.

Some of these dreads seem to be purely physical in origin or character yet prove to be merely or to a great degree only psychic states. Insomnia itself is more a dread than anything else. In writing for the International Clinics some years ago (Volume IV, Series XXVI) I dwelt on the fact that insomnia as a dread was probably responsible for more discomfort and complaints from mankind than almost anything else. Insomniaphobia is just such a dread as agoraphobia, the dread of open spaces; or akrophobia, the dread of heights; or skotophobia, the dread of the dark, and other phobias which afflict mankind. It is perfectly possible in most cases to cure such phobias by direct training against them, and this can be done also with regard to insomnia.

Some people, particularly those who have not been out much during the day and who have suffered from wakefulness a few times, get it on their mind that if this state keeps up they will surely lose their reason or their bodily health, and they begin to worry about it. They commence wondering about five in the afternoon whether they are going to be awake that night or not. It becomes a haunt, and no matter what they do during the evening every now and then the thought recurs that they will not sleep. By the time they actually lie down they have become so thoroughly occupied with that thought that it serves to keep them awake. Some of them avoid the solicitude before they actually get to bed, but begin to worry after that, and if after ten minutes they are not asleep, above all if they hear a clock strike somewhere, they are sure they are going to be awake, they worry about it, get themselves thoroughly aroused, and then they will not go to sleep for hours. It is quite useless to give such people drugs, just as useless as to attempt to give a man a drug to overcome the dread of heights or the dread of the dark or of a narrow street through which he has to pass. They must use their wills to help them out of a condition in which their dreads have placed them.

Apart from these neurotic dreads, quite unreasoning as most of them are, there are a series of what may be called intellectual dreads. These are due to false notions that have come to be accepted and that serve to keep people from doing things that they ought to do for the sake of their health, or set them performing acts that are injurious instead of beneficial. The dread of loss of sleep has often caused people to take somnifacients which eventually proved ever so much more harmful than would the loss of sleep they were meant to overcome. Many a person dreading a cold has taken enough quinine and whisky to make him more miserable the next day than the cold would have, had it actually made its appearance, as it often does not. The quinine and whisky did not prevent it, but the expectation was founded on false premises. There are a great many other floating ideas that prove the source of disturbing dreads for many people. A discussion of a few typical examples will show how much health may be broken by the dreads associated with various ills, for they often interfere with normal, healthy living.

"A little knowledge is a dangerous thing" applies particularly in this matter. There are many morbid fears that disturb mankind and keep us from accomplishing what might otherwise be comparatively easy. A great many people become convinced that they have some diseased condition, or morbid elements at least, in them which make it impossible for them to do as much as other people. Sometimes this morbid persuasion takes the form of hypochrondia and the individuals feel that they have a constitution that unfits them for prolonged and strenuous effort of any kind, so they avoid it. The number of valetudinarians, that is of those who live their lives mainly engaged in caring for their health, though their physicians have never been able to find anything organically wrong with them, is much larger than might be imagined. This state of mind has been with us for many centuries, for the word which describes it, hypochondria, came to us originally from Greece and is an attempt to localize the affection in connection with its principal symptom, which is usually one of discomfort in the stomach region or to one side or the other of it, that is, in the hypochondria or beneath the ribs.

Such a state of mind, in which the patient is constantly complaining of one symptom or another, quite paralyzes the will. The individual may be able to do some routine work but he will not be able to have any initiative or energy for special developments of his occupation, and of course, when any real affection occurs, he will feel that he is quite unable to bear this additional burden of disease. Hypochondriacs, however, sometimes fairly enjoy their ill health and therefore have been known not infrequently to live on to a good, round old age, ever complaining more and more. It is their dread of disease that keeps them from getting better and prevents their wills from throwing off whatever symptoms there are and becoming perfectly well. Until something comes along and rouses their wills, there is no hope of affecting them favorably, and it is surprising how long the state may continue without any one ever having found any organic affection to justify all the discomforts of which they complain. Quite literally, they are suffering from complaints and not from disease in the ordinary sense of the word.

Sometimes these dreads of disease are dependent on some word which has taken on an exaggerated significance in people's minds. A word that in recent years has been the source of a great deal of unfavorable suggestion is "catarrh", and a mistaken notion of its meaning has been productive of a serious hampering of their will to be well in a number of persons. In itself, both according to its derivation and its accepted scientific significance, the word means only that first stage of inflammatory irritation of mucous membranes which causes secretion to flow more freely than normally. Catarrhein in Greek means only to flow down. [Footnote]

[Footnote : The word has, by the way, the same meaning as rheumatism, which is also from the Greek verb, to flow, though its application is usually limited to the serous membranes of the joints or the serous surfaces of the intermuscular planes. By derivation, catarrh is the same word also as gout, which comes from gutta in Latin, meaning a drop and implying secretory disturbances. These three words catarrh, rheumatism, gout have been applied to all sorts of affections and are so general in meaning as to be quite hard to define exactly. They have for this very reason, their vagueness, become a prolific source of unfortunate suggestion and of all kinds of dreads that disturb health.]

By abuse, however, the word catarrh has come to mean in the minds of a great many people in our time a very serious inflammation of the mucous membranes, almost inevitably progressive and very often resulting in fetid diseased conditions of internal or external mucous membranes, very unpleasant for the patient and his friends and the source of serious complications and sequelae. This idea has been fostered sedulously by the advertisers of proprietary remedies and the ingenious exploiters of various modes of treatment. As a result, a great many people who for one reason or another usually because of some slight increase of secretion in the nose and throat become convinced they have catarrh begin to feel that they cannot be expected to have as much resistive vitality as others, since they are the subjects of this serious progressive disease. As a matter of fact, very few people in America, especially those living in the northern or eastern States, are without some tendency to mild chronic catarrh. The violent changes of temperature and the damp, dark days predispose to it; but it produces very few symptoms except in certain particularly sensitive individuals whose minds become centered on slight discomforts in the throat and nose and who feel that they must represent some serious and probably progressive condition.

As a matter of fact, catarrh has almost nothing of the significance attributed to it so often in magazine and newspaper advertisements. Simple catarrh decreases without producing any serious result, and indeed it is an index of a purely catarrhal condition that there is a complete return to normal. Sometimes microbes are associated with its causation, but when this is so, they are bacteria of mild pathological virulence that do not produce deep changes. As for catarrh developing fetid, foul-smelling discharges or odors, that is out of the question. There are certain affections, notably diphtheria, that may produce such serious changes in the mucous membranes that there will always even long after complete recovery be an unpleasant odorous condition, but it is probable that even in these cases there exists a special form of microbe quite rare in occurrence which produces the state known as ozena.

As to catarrh spreading from the nose and throat to the other mucous membranes, that is also quite out of the question if it is supposed to occur in the way that the advertising specialist likes to announce. Catarrhal conditions may occur in the stomach, but like those of the nose and throat they are not serious, heal completely, and produce no definite changes. A pinch of snuff may cause a catarrhal condition of the nose, that is an increase of secretion due to hyperaemia of the mucous membrane; the eating of condiments, of Worcestershire sauce, peppers, and horse-radish may cause it in the stomach. It may be due to microbic action or to irritant or decomposing food, but it is not a part of a serious, wide-spreading pathological condition that will finally make the patient miserable. It is surprising, however, how many people say with an air of finality that they have catarrh, as if it should be perfectly clear that as a result they cannot be expected at any time to be in sufficiently good health to be called on for any special work, and of course if any affection should attack them, their natural immunity to disease has been so lowered by this chronic affection, of which they are the victims, that no strong resistance could be expected from them.

All this is merely a dread induced by paying too much attention to medical advertisements. It is better not to know as much as some people know, or think they know about themselves, than to know so many things that are not so. Their dreads seriously impair their power to work and leave them ill disposed to resist affections of any kind that may attack them. It is a sad confession to make, but not a little of the enforced study of physiology in our schools has become the source of a series of dreads and solicitudes rather than of helpful knowledge. We have as a result a generation who know a little about their internal economy, but only enough to make them worry about it and not quite enough to make them understand how thoroughly capable our organisms are of caring for themselves successfully and with resultant good health, if we will only refrain from putting brakes on their energies and disturbing their functions by our worries and anxieties.

Another such word as catarrh in its unfavorable suggestiveness in recent years has been auto-intoxication. It is a mouth-filling word, and therefore very probably it has occupied the minds of the better educated classes. Usually the form of auto-intoxication that is most spoken of is intestinal auto-intoxication, and this combination has for many people a very satisfying polysyllabic length that makes it of special significance. Its meaning is taken to be that whenever the contents of the intestines are delayed more than twenty hours or perhaps a little longer, or whenever certain irritant materials find their way into the intestinal tract, there is an absorption of toxic matter which produces a series of constitutional symptoms. These include such vague symptomatic conditions as sleepiness, torpor after meals, an uncomfortable sense of fullness though when we were young we rather liked to have that feeling of fullness and sometimes a feeling of heat in the skin with other sensations of discomfort in various parts of the body. At times there is headache, but this is rather rare; lassitude and a feeling of inability to do things is looked upon as almost characteristic of the condition. Usually there are nervous symptoms of one kind or another associated with the other complaints and there may be distinctly hysterical or psycho-neurotic manifestations.

Auto-intoxication as just described has become a sort of fetish for a great many people who bow down and worship at its shrine and give some of the best of their energies and not a little of their time to meditation before it. As a matter of fact, in the last few years it has come to be recognized that auto-intoxication is a much abused word employed very often when there are serious organic conditions in existence elsewhere in the body and still more frequently when the symptoms are due merely to functional nervous troubles. These are usually consequent upon a sedentary life, lack of fresh air and exercise, insufficient attention to the diet in the direction of taking simple and coarse food, and generally passing disturbances that can be rather readily catalogued under much simpler affections than a supposed absorption of toxic materials from the intestines. Reflexes from the intestinal tract, emphasized by worries about the condition, are much more responsible for the feelings complained of which are often not in any sense symptoms than any physical factors present.

As Doctor Walter C. Alvarez said in a paper on the "Origin of the So-called Auto-intoxicational Symptoms" published from the George Williams Hooper Foundation for Medical Research of the University of California Medical School, [Footnote] as the conclusion of his investigation of the subject:

"Auto-intoxication is commonly diagnosed when a physical examination would show other more definite causes for the symptoms. Those who believe that intestinal stasis can account for a long list of disease conditions have little proof to offer for their views. Many of the assumptions on which they rest their case have proved to be wrong.

"The usual symptoms of the constipated disappear so promptly after a bowel movement that they cannot be due to absorbed toxins. They must be produced mechanically by distension and irritation of the colon. They occur in nervous, sensitive people. It has been shown that various activities of the digestive tract can profoundly affect the sensorium and the vasomotor nerves. The old ideas of insidious poisoning led to the formation of hypochrondriacs; the new explanation helps to cure many of them."

There are many other terms in common use that have unfortunate suggestions and make people feel, if they once get the habit of applying them to themselves, that they are the subject of rather serious illness. I suppose that one of the most used and most abused of these is uric acid and the uric acid diathesis. Scientific physicians have nearly given up these terms, but a great many people are still intent on making themselves miserable. All sorts of symptoms usually due to insufficient exercise and air, inadequate diversion of mind and lack of interests are attributed to these conditions. Some time or other a physician or perhaps some one who is supposed to be a friend suggested them and they continue to hamper the will to be well by baseless worries founded on false notions for years afterwards. What is needed is a definite effort of the will to throw off these nightmares of disease that are so disturbing and live without them.

It is surprising how much vital energy may be wasted in connection with such dreads. Unfortunately, too, medicines of various kinds are taken to relieve the symptoms connected with them and the medicine does ever so much more harm than good. Oliver Wendell Holmes declared a generation ago that if all the medicines that had ever been taken by mankind were thrown into the sea it would be much better for mankind and much worse for the fishes. The expression still has a great truth in it, especially as regards that habit of self-drugging so common among the American people. In the course of lecture engagements, I stay with very intelligent friends on a good many occasions each year, and it is surprising how many of them have medicine bottles around, indicating that they are subject to dreads of various kinds with regard to themselves for which they feel medicine should be taken. These dreads unfortunately often serve to lessen resistive vitality to real affections when they occur and therefore become a source of real danger.

All these various dreads, then, have the definite effect of lessening the power of the will to enable people to do their work and remain well. They represent serious brakes upon the flow of nerve impulses from the spiritual side of man's nature to the physical. This is much more serious in its results than would usually be thought; and one of the things that a physician has to find out from a great many patients is what sources of dread they are laboring under so as to neutralize them or at least correct them as far as possible. It is surprising how much good can be accomplished by a deliberate quest after dreads and the direct discussion of them, for they are always much less significant when brought out of the purlieus of the mind directly into the open. Many a neurotic patient, particularly, will not be improved until his dreads are relieved. This form of psycho-analysis rather than the search for sex insults, as they are called, or sexual incidents of early life, is the hopeful phase of modern psychological contribution to therapeutics.

CHAPTER III

HABITS

"Why, will shall break it; will and nothing else."
Love's Labor's Lost

Dreads are brakes on the will, inhibitions which prevent its exercise and make accomplishment very difficult and sometimes impossible. They represent mainly a state of mind, yet often they contain physical elements, and the disposition counts for much. Their counterpart in the opposite direction is represented by habits which are acquired facilities of action for good or for ill. Habits not only make activities easy but they even produce such a definite tendency to the performance of certain actions as to make it difficult not to do them. They may become so strong as to be tyrants for ill, though it must not be forgotten that properly directed they may master what is worst in us and help us up the hill of life. Acts that are entirely voluntary and very difficult at first may become by habit so natural that it is extremely difficult to do otherwise than follow the ingrained tendency. Nature's activities are imperative. Habitual actions may become equally so. When some one once remarked to the Duke of Wellington that habit was second nature, he replied:

"Oh, ever so much more than that! Habit may be ten times as strong as nature."

The function of the will in health is mainly to prevent the formation of bad habits or break those that have been formed, but above all, to bring about the formation of habits that will prevent as far as is possible the development of tendencies to disease in the body, Man probably faces no more difficult problem in life than the breaking of a bad habit. Usually it requires the exercise of all his will power applied to its fullest extent. If there is a more difficult problem than the breaking of a bad habit it is the formation of a good one late in life because of the persistency of advertence and effort that is required. It is comparatively easy to prevent the formation of bad habits and also easy to form good habits in the earlier years. The organism is then plastic and yields itself readily and thus becomes grooved to the habit or hardened against it by the performance of even a few acts.

All the psychologists insist that after the period of the exercise of instinct as the basis of life passes, habit becomes the great force for good or for ill. We become quite literally a bundle of habits, and the success of life largely depends on whether these habits are favorable or unfavorable to the accomplishment of what is best in us. More than anything else health depends on habit. We begin by doing things more or less casually, and after a time a tendency to do them is created; then almost before we know it, we find that we have a difficult task before us, if we try not to do them.

To begin with, the activity which becomes the subject of a habit may be distinctly unpleasant and require considerable effort to accomplish. Practically every one who has learned to smoke recalls more or less vividly the physical disturbance caused by the first attempt and how even succeeding smokes for some time, far from being pleasant, required distinct effort and no little self-control. After a time, the desire to smoke becomes so ingrained that a man is literally made quite miserable by the lack of it and finds himself almost incapable of doing anything else until he has had his smoke.

Even more of an effort is required to establish the habit of chewing tobacco, and it is even more difficult to break when once it has been formed. Any one who has seen the discomfort and even torments endured by a man who, after he had chewed tobacco for many years, tried to stop will appreciate fully what a firm hold the habit has obtained. I have known a serious business man who almost had to give up business, who lost his sleep and his appetite and went through a nervous crisis merely by trying to break the habit of chewing tobacco.

In the Orient they chew betel nut. It is an extremely hot material which burns the tongue and which a man can stand for only a very short time when he first tries it. After a while, however, he finds a pleasant stimulation of sensation in the constant presence of the biting betel nut in his mouth; he craves it and cannot do his work so well without it. He will ever advert to its use and will be restless without it. He continues to use it in spite of the fact that the intense irritation set up by the biting qualities of the substance causes cancer of the tongue to occur ten times as frequently among those who chew betel nut as among the rest of the population. Not all those who chew it get cancer, for some die from other causes before there is time for the cancer to develop, and some seem to possess immunity against the irritation. The betel nut chewer ignores all this, proceeds to form the habit, urged thereto by the force of example, and then lets himself drift along, hoping that it will have no bad effects.

The alcohol and drug habits are quite as significant in shortening life as betel nut and yet men take them up quite confident in the beginning that they will not fall victims, and then find themselves enmeshed. It is probable that the direct physical effects of none of these substances shorten life to a marked degree unless they are indulged in to very great excess, but the moral hazards which they produce, accidents, injuries of various kinds, exposure to disease, all these shorten life. Men know this very well, and yet persist in the formation of these habits.

Any habit, no matter how strong, can be broken if the individual really wishes to break it, provided the subject of it is not actually insane or on the way to the insane asylum. He need only get a motive strong enough to rouse his will, secure a consciousness of his own power, and then the habit can be broken. After all, it must never be forgotten that the only thing necessary in order to break a habit effectively is to refuse to perform a single act of it, the next time one is tempted. That breaks the habit and makes refusal easier and one need only continue the refusal until the temptation ceases.

Men who have not drawn a sober breath for years have sometimes come to the realization of the fools that they were making of themselves, the injury they were doing their relatives, or perhaps have been touched by a child's words or some religious motive, and after that they have never touched liquor again. Father Theobald Mathew's wonderful work in this regard among the Irish in the first half of the nineteenth century has been repeated by many temperance or total abstinence advocates in more recent generations. I have known a confirmed drunkard reason himself into a state of mind from which he was able to overcome his habit very successfully, though his reasoning consisted of nothing more than the recognition of the fact that suggestion was the root of his craving for alcohol. His father had been a drunkard and he had received so many warnings from all his older relatives and had himself so come to dwell on the possible danger of his own formation of the habit that he had suggested himself into the frame of mind in which he took to drink. I have known a physician on whom some half a dozen different morphine cures had been tried always followed by a relapse cure himself by an act of his own will and stay cured ever since because of an incident that stirred him deeply enough to arouse his will properly to activity. One day his little boy of about four was in his office when father prepared to give himself one of his usual injections of morphine. The little boy gave very close attention to all his father's manipulations, and as the doctor was hurrying to keep an appointment, he did not notice the intent eye witness of the proceedings. Just as the needle was pushed home and the piston shot down in the barrel, the little boy rushed over to his father and said, "Oh, Daddy, do that to me." Apparently this close childish observer had noted something of the look of satisfaction that came over his father's face as he felt the fluid sink into his tissues. It is almost needless to say that the shock the father received was enough to break his morphine habit for good and all. It simply released his will and then he found that if he really wanted to, he could accomplish what the various cures for the morphine habit only lead up to and in his case unsuccessfully the exercise of his own will power.

The word "habit" suggests nearly always, unfortunately, the thought of bad habits, just as the word "passion" implies, with many people, evil tendencies. But it must not be forgotten that there are good passions and good habits that are as helpful for the accomplishment of what is best in life as bad passions and bad habits are harmful. A repetition of acts is needed for the formation of good habits just as for the establishment of customs of evil. Usually, however, and this must not be forgotten, the beginning of a good habit is easier than the beginning of a bad habit. Once formed, the good habits are even more beneficial than the bad habits are harmful. It is almost as hard to break a good habit as a bad one, provided that it has been continued for a sufficient length of time to make that groove in the nervous system which underlies all habit. We cannot avoid forming habits and the question is, shall we form good or bad habits? Good habits preserve health, make life easier and happier; bad habits have the opposite effect, though there is some countervailing personal element that tempts to their formation and persistence.

Every failure to do what we should has its unfortunate effect upon us. We get into a state in which it is extremely difficult for us to do the right things. We have to overcome not only the original inertia of nature, but also a contrary habit. If we do not follow our good impulses, the worse ones get the upper hand. As Professor James said, for we must always recur to him when we want to have the clear expression of many of these ideas:

"Just as, if we let our emotions evaporate, they get into a way of evaporating; so there is reason to suppose that if we often flinch from making an effort, before we know it the effort-making capacity will be gone; and that, if we suffer the wandering of our attention, presently it will wander all the time. Attention and effort are but two names for the same psychic fact. To what brain-processes they correspond we do not know. The strongest reason for believing that they do depend on brain processes at all and are not pure acts of the spirit, is just this fact, that they seem in some degree subject to the law of habit, which is a material law."

It must not be forgotten that we mold not alone what we call character, but that we manifestly produce effects upon our tissues that are lasting. Indeed it is these that count the most, for health at least. It is the physical basis of will and intellect that is grooved by what we call habit. As Doctor Carpenter says:

"Our nervous systems have grown to the way in which they have been exercised, just as a sheet of paper or a coat, once creased or folded, tends to fall forever afterwards into the same identical fold."

Permitting exceptions to occur when we are forming a habit is almost necessarily disturbing. The classical figure is that it is like letting fall a ball of string which we have been winding. It undoes in a moment all that we have accomplished in a long while. As Professor Bain has said it so much better than I could, I prefer to quote him:

"The peculiarity of the moral habits, contradistinguishing them from the intellectual acquisitions, is the presence of two hostile powers, one to be gradually raised into the ascendant over the other. It is necessary, above all things, in such a situation never to lose a battle. Every gain on the wrong side undoes the effect of many conquests on the right. The essential precaution, therefore, is so to regulate the two opposing powers that the one may have a series of uninterrupted successes, until repetition has fortified it to such a degree as to enable it to cope with the opposition under any circumstances."

This means training the will by a series of difficult acts, accomplished in spite of the effort they require, but which gradually become easier from repeated performance until habit replaces nature and dominates the situation.

Serious thinkers who faced humanity's problems squarely and devoted themselves to finding solutions for them had worked out this formula of the need of will training long ago, and it was indeed a principal characteristic of medieval education. The old monastic schools were founded on the idea that training of the will and the formation of good habits was ever so much more important than the accumulation of information. They frankly called the human will the highest faculty of mankind and felt that to neglect it would be a serious defect in education. The will can only be trained by the accomplishment of difficult things day after day until its energies are aroused and the man becomes conscious of his own powers and the ability to use them whenever he really wishes. There was a time not so long since, and there are still voices raised to that purport, when it was the custom to scoff at the will training of the older time and above all the old-fashioned suggestion that mortifications of various kinds that is, the doing of unpleasant things just for the sake of doing them should be practiced because of the added will power thus acquired. The failure of our modern education which neglected this special attention to the will is now so patent as to make everyone feel that there must be a recurrence to old time ideas once more.

The formation of proper habits should, then, be the main occupation of the early years. This will assure health as well as happiness, barring the accidents that may come to any human being. Good habits make proper living easy and after a time even pleasant, though there may have been considerable difficulty in the performance of the acts associated with them at the beginning. Indeed, the organism becomes so accustomed to their performance after a time that it becomes actually something of a trial to omit them, and they are missed.

Education consists much more in such training of the will than in storing the intellect with knowledge, though the latter idea has been unfortunately the almost exclusive policy in our education in recent generations. We are waking up to the fact that diminution of power has been brought about by striving for information instead of for the increase of will energy.

Professor Conklin of Princeton, in his volume on "Heredity and Environment", emphasized the fact that "Will is indeed the supreme human faculty, the whole mind in action, the internal stimulus which may call forth all the capacities and powers." He had said just before this: "It is one of the most serious indictments against modern systems of education that they devote so much attention to the training of the memory and intellect and so little attention to the training of the will, upon the proper development of which so much depends."

Nor must it be thought that the idea behind this training of the will is in any sense medievally ascetic and old-fashioned and that it does not apply to our modern conditions and modes of thinking. Professor Huxley would surely be the one man above all whom any one in our times would be least likely to think of as mystical in his ways or medieval in his tendencies. In his address on "A Liberal Education and Where to Find It", delivered before the South London Workingmen's College some forty years ago, in emphasizing what he thought was the real purpose of education, he dwelt particularly on the training of the will. He defined a liberal education not as so many people might think of it in terms of the intellect, but rather in terms of the will. He said that a liberal education was one "which has not only prepared a man to escape the great evils of disobedience to natural laws, but has trained him to appreciate and to seize upon the rewards which nature scatters with as free a hand as her penalties." And then he added:

"That man, I think, has had a liberal education who has been so trained in youth that his body is the ready servant of his will, and does with ease and pleasure all the work that, as a mechanism, it is capable of; whose intellect is a clear, cold, logic engine, with all its parts of equal strength, and in smooth working order, ready, like a steam engine, to be turned to any kind of work, and spin the gossamers as well as forge the anchors of the mind; whose mind is stored with a knowledge of the great and fundamental truths of nature and of the laws of her operations; one who is no stunted ascetic but who is full of life and fire, but whose passions are trained to come to heel by a vigorous will, the servant of a tender conscience; who has learned to love all beauty, whether of nature or of art, to hate all vileness, and to respect others as himself.

"Such an one and no other, I conceive, has had a liberal education; for he is, completely as a man can be, in harmony with nature."

This is the liberal education in habits of order and power that every one must strive for, so that all possible energies may be available for the rewards of good health. Details of the habits that mean much for health must be reserved for subsequent chapters, but it must be appreciated in any consideration of the relation of the will to health that good habits formed as early as possible in life and maintained conservatively as the years advance are the mainstay of health and the power to do work.

CHAPTER IV

SYMPATHY

"Never could maintain his part but in the force of his will."
Much Ado about Nothing

A great French physician once combined in the same sentence two expressions that to most people of the modern time would seem utter paradoxes. "Rest," he said, "is the most dangerous of remedies, never to be employed for the treatment of disease, except in careful doses, under the direction of a physician and rarely for any but sufferers from organic disease"; while "sympathy", he added, "is the most insidiously harmful of anodynes, seldom doing any good except for the passing moment, and often working a deal of harm to the patient."

With the first of these expressions, we have nothing to do here, but the second is extremely important in any consideration of the place of the will in human life. Nothing is so prone to weaken the will, to keep it from exerting its full influence in maintaining vital resistance, and as a result, to relax not only the moral but the physical fiber of men and women as misplaced sympathy. It has almost exactly the same place in the moral life that narcotics have in the physical, and it must be employed with quite as much nicety of judgment and discrimination.

Sympathy of itself is a beautiful thing in so far as it implies that suffering with another which its Greek etymology signifies. In so far as it is pity, however, it tends to lessen our power to stand up firmly under the trials that are sure to come, and is just to that extent harmful rather than helpful. There is a definite reaction against it in all normal individuals. No one wants to be pitied. We feel naturally a little degraded by it. In so far as it creates a feeling of self-pity, it is particularly to be deprecated, and indeed this is so important a subject in all that concerns the will to be well and to get well that it has been reserved for a special chapter. What we would emphasize here is the harm that is almost invariably done by the well-meant but so often ill-directed sympathy of friends and relatives which proves relaxing of moral purpose and hampers the will in its activities, physical as well as ethical.

Human nature has long recognized this and has organized certain customs of life with due reference to it. We all know that when children fall and even hurt themselves, the thing to do is not to express our sympathy and sorrow for them, even though we feel it deeply, but unless their injury is severe, to let them pick themselves up and divert their minds from their hurts by suggesting that they have broken the floor, or hurt it. For the less sympathy expressed, the shorter will be the crying, and the sooner the child will learn to take the hard knocks of life without feeling that it is especially abused or suffering any more than comes to most people. Unfortunately, it is not always the custom to do the same thing with the children of a larger growth. This is particularly true when there is but a single child in the family, or perhaps two, when a good deal of sympathy is likely to be wasted on their ills which are often greatly increased by their self-consciousness and their dwelling on them. Diversion of mind, not pity, is needed. The advice to do the next thing and not cry over spilt milk is ever so much better than sentimental recalling of the past.

Many a young man who went to war learned the precious lesson that sympathy, though he might crave it, instead of doing him good would do harm. Many a manly character was rounded out into firm self-control and independence by military discipline and the lack of anything like sentimentality in camp and military life. A good many mothers whose boys had been the objects of their special solicitude felt very sorry to think that they would have to submit to the hardships and trials involved in military discipline. Most of them who were solicitous in this way were rather inclined to feel that their boy might not be able to stand up under the rigidities of military life and hoped at most that he would not be seriously harmed. They could not think that early rising, hard work, severe physical tasks, tiring almost to exhaustion, with plain, hearty, yet rather coarse food, eaten in slapdash fashion, would be quite the thing for their boy of whom they had taken so much care. Not a few of them were surprised to find how the life under these difficult circumstances proved practically always beneficial.

I remember distinctly that when the soldiers were sent to the Mexican border the mother of a soldier from a neighboring State remarked rather anxiously to me that she did not know what would happen to Jack under the severe discipline incident to military life. He had always gone away for five or six weeks in the summer either to the mountains or to the seashore, and the Mexican border, probably the most trying summer climate in the United States, represented the very opposite of this. Besides, there was the question of the army rations; Jack was an only son with five sisters. Most of them were older than he, and so Jack had been coddled as though by half a dozen mothers. He was underweight, he had a rather finicky appetite, he was capricious in his eating both as to quantity and quality, and was supposed to be a sufferer from some form of nervous indigestion. Personally, I felt that what Jack needed was weight, but I had found it very hard to increase his weight. He was particularly prone to eat a very small breakfast, and his mother once told me that whenever he was at home, she always prepared his breakfast for him with her own hands. This did not improve matters much, however, for Jack was likely to take a small portion of the meat cooked for him, refuse to touch the potatoes, and eat marmalade and toast with his coffee and nothing more. No wonder that he was twenty pounds underweight or that his mother should be solicitous as to what might happen to her Jack in army life at the Border.

I agreed with her in that but there were some things that I knew would not happen to Jack. His breakfast, for instance, would not be particularly cooked for him, and he might take or leave exactly what was prepared for every one else. Neither would the Government cook come out and sit beside Jack while he was at breakfast and tempt him to eat, as his mother had always done. I knew, too, that at other meals, while the food would be abundant, it would usually be rather coarse, always plain, and there would be nothing very tempting about it unless you had your appetite with you. If ever there is a place where appetite is the best sauce, it is surely where one is served with army food.

I need scarcely tell what actually happened to Jack, for it was exactly what happened to a good many Jacks whose mothers were equally afraid of the effects of camp life on them. Amid the temptations of home food, Jack had remained persistently underweight. Eating an army ration with the sauce of appetite due to prolonged physical efforts in the outdoor air every day, Jack gained more than twenty pounds in weight, in spite of the supposedly insalubrious climate of the Border and the difficult conditions under which he had to live. It was literally the best summer vacation that Jack had ever spent, though if the suggestion had ever been made that this was the sort of summer vacation that would do him good, the idea would have been scoffed at as impractical, if not absolutely impossible.

Homer suggested that a mollycoddle character whom he introduces into the "Iliad" owed something of his lack of manly stamina to the fact that he had six sisters at home, and an Irish friend once translated the passage by saying that the young man in question was "one of seven sisters." This had been something of Jack's trouble. He had been asked always whether he changed his underwear at the different seasons, whether he wore the wristlets that sisterly care provided for him, whether he put on his rubbers when he went out in damp weather and carried his umbrella when it was threatening rain, and all the rest. He got away from all this sympathetic solicitude in army life and was ever so much better for it.

It is extremely difficult to draw the line where the sympathy that is helpful because it is encouraging ends, and sentimental pity which discourages begins. There is always danger of overdoing and it is extremely important that growing young folks particularly should be allowed to bear their ills without help and learn to find resources within themselves that will support them. The will can thus be buttressed to withstand the difficulties of life, make them much easier to bear, and actually lessen their effect. Ten growing young folks have been seriously hurt by ill-judged sympathy for every one that has been discouraged by the absence of sympathy or by being made to feel that he must take the things of life as they come and stand them without grouchy complaint or without looking for sympathy.

This is particularly true as regards those with any nervous or hysterical tendencies, for they readily learn to look for sympathy. The most precious lesson of the war for physicians has been that which is emphasized in the chapter on "The Will and the War Psychoneuroses." There was an immense amount of so-called "shell-shock" which really represented functional neurotic conditions such as in women used to be called hysteria. At the beginning of the war there was a good deal of hearty sympathy with it, and patients were encouraged by the physicians and then by the nurses and other patients in the hospital to tell over and over again how their condition developed. It was found after a time that the sympathy thus manifested always did harm. The frequent repetition of their stories added more and more suggestive elements to the patients' condition, and they grew worse instead of better. It was found that the proper curative treatment was to make just as little as possible of their condition, to treat them firmly but with assurance once it had been definitely determined that no organic nervous trouble was present and to bring about a cure of whatever symptoms they had at a single sitting by changing their attitude of mind towards themselves.

Some of the patients proved refractory and for these isolation and rather severe discipline were occasionally necessary. The isolation was so complete as to deprive them not only of companionship but also of reading and writing materials and the solace of their tobacco. Severe cases were sometimes treated by strong faradic currents of electricity which were extremely painful. Patients who insisted that they could not move their muscles were simply made to jump by an electric shock, thus proving to them that they could use the muscles, and then they were required to continue their use.

Those suffering from shell-shock deafness and muteness were told that an electrode would be applied to their larynx or the neighborhood of their ear and when they felt pain from it, that was a sign that they were able to talk and to hear if they wished, and that they must do so. Relapses had to be guarded against by suggestion, and where relapses became refractory and stronger currents of electricity to ear and larynx were deemed inadvisable, the strict isolation treatment usually proved effective.

In a word, discipline and not sympathy was the valuable mode of treating them. Sympathy did them harm as it invariably does. The world has recognized this truism always, but we need to learn the lesson afresh, or the will power is undermined. Character is built up by standing the difficult things of life without looking for the narcotic of sympathy or any other anaesthetizing material. These are "hard sayings," to use a Scriptural expression, but they represent the accumulation of wisdom of human experience. Sympathy can be almost as destructive of individual morale as the dreads, and it is extremely important that it should not be allowed to sap will energy. In our time above all, when the training of the will has been neglected, though it is by far the most important factor in education, this lesson with regard to the harmful effect of sympathy needs to be emphasized.

For nervous people, that is, for those who have, either from inheritance or so much oftener from environment, yielded to circumstances rather than properly opposed them, sympathy is quite as dangerous as opium. George Eliot once replied to a friend who asked her what was duty, that duty consisted in facing the hard things in life without taking opium.

Healthy living to a great extent depends on standing what has to be borne from the bodies that we carry around with us without looking for sympathy. It has often been emphasized that human beings are eminently lonely. The great experiences of life and above all, death and suffering, we have to face by ourselves and no one can help us. We may not be, as Emerson suggested, "infinitely repellent particles", but at all the profoundest moments of life we feel our alone-ness. The more that we learn to depend bravely on ourselves and the less we seek outside support for our characters, the better for us and our power to stand whatever comes to us in life.

Physical ills are always lessened by courageously facing them and are always increased by cringing before them. The one who dreads suffers both before and during the time of the pain and thus doubles his discomfort. We must stand alone in the matter and sympathy is prone to unman us. Looking for sympathy is a tendency to that self-pity which is treated in a subsequent chapter and which does more to increase discomfort in illness, exaggerate symptoms, and lower resistive vitality than anything else, in the psychic order at least.

Suffering is always either constructive or destructive of character. It is constructive when the personal reaction suffices to lessen and make it bearable. It is destructive whenever there is a looking for sympathy or a leaning on some one else. Character counts in withstanding disease, and even in the midst of epidemics, according to many well-grounded traditions, those who are afraid contract the disease sooner than others and usually suffer more severely. Sympathy must not be allowed to produce any such effect as this.

CHAPTER V

SELF-PITY

"The will dotes that is attributive
To what infectiously itself affects."
Troilus and Cressida

The worst brake on the will to be well is undoubtedly the habit that some people have of pitying themselves and feeling that they are eminently deserving of the pity of others because of the trials, real or supposed, which they have to undergo. Instead of realizing how much better off they are than the great majority of people for most of the typical self-pitiers are not real subjects for pity they keep looking at those whom they fondly suppose to be happier than themselves and then proceed to get into a mood of commiseration with themselves because of their ill health real or imaginary or uncomfortable surroundings. Just as soon as men or women assume this state of mind, it becomes extremely difficult for them to stand any real trials that appear, and above all, it becomes even more difficult for them to react properly against the affections of one kind or another that are almost sure to come. Self-pity is ever a serious hamperer of resistive vitality.

A great many things in modern life have distinctly encouraged this practice of self-pity and conscious commiseration of one's state until it has become almost a commonplace of modern life for those who feel that they are suffering, especially if they belong to what may be called the sophisticated classes. We have become extremely sensitive as a consequence about contact with suffering. Editors of magazines and readers for publishing houses often refuse in our time to accept stories that have unhappy endings, because people do not care to read them, it is said. The story may have some suffering in it and even severe hardships, especially if these can be used for purposes of dramatic climax, but by the end of the story everything must have turned out "just lovely", and it must be understood that suffering is only a passing matter and merely a somewhat unpleasant prelude to inevitable happiness.

Almost needless to say, this is not the way of life as it must be lived in what many generations of men have agreed in calling "this vale of tears." For a great many people have to suffer severely and without any prospect of relief none of us quite escape the necessity of suffering and as some one has said, all human life, inasmuch as there is death in it, must be considered a tragedy. The old Greeks did not hesitate, in spite of their deep appreciation of the beauty of nature and cordial enthusiasm for the joy of living, even to emphasize the tragedy in life. They were perhaps inclined to think that the sense of contrast produced by tragedy heightened the actual enjoyment of life and that indeed all pleasure was founded rather on contrast than positive enjoyment. One may not be ready to agree with the saying that the only thing that makes life worth while is contrast, but certainly suffering as a background enhances happiness as nothing else can.

Aristotle declared that tragedy purges life, that is, that only through the lens of death and misfortune could one see life free from the dross of the sordid and merely material to which it was attached. His meaning was that tragedy lifted man above the selfishness of mere individualism, and by showing him the misfortunes of others prepared him to struggle for himself when misfortune might come, as it almost inevitably would; and at the same time lifted him above the trifles of daily life into a higher, broader sphere of living, where he better realized himself and his powers.

For man is distinctly prone to forget about death and suffering, and when he does, to become eminently selfish and forgetful of the rights of others and his duties towards them. The French have a saying, consisting of but four words and an intervening shrug of the shoulders, that is extremely illuminating. They quote as the expression of the usual thought of men when brought face to face with the fact that people are dying all around them, "On meurt les autres!" "People die Oh, yes (with an expressive shrug of the shoulders), other people!" We refuse to recognize the fact that we too must go until that is actually forced upon us by advancing years or by some incurable disease. As for suffering, a great many people have come almost to resent that they should be asked to suffer, and character dissolves in self-pity as a result.

Instead of the constant, continuous reading of what may be called Sybaritic literature for it is said that the Sybarite finds it impossible to sleep if there is a crushed rose leaf next his skin instead of being absorbed in the literature which emphasizes the pleasures of life and pushes its pains into the background, young people, and especially those of the better-to-do classes, should be taught from their early years to read the lives of those who have endured successfully hardships of various kinds and have succeeded in getting satisfaction out of their accomplishment in life, despite all the suffering that was involved. These are human beings like ourselves, and what mortal has done, other mortals can do.

There was a school of American psychologists before the war who had come to recognize the value of that old-fashioned means of self-discipline of mind, the reading of the lives of the saints. For those to whom that old-fashioned practice may seem too reactionary, there are the lives and adventures of our African and Asiatic travelers and our polar explorers as a resource.

War books have been a godsend for our generation in this regard. They have led people to contemplate the hardest kind of suffering and very often in connection with those who are nearest and dearest to them and thus made them understand something of the possibilities of human nature to withstand trials and sufferings. As a result they have been trained not to make too much of their own trivial trials, as they soon learned to recognize them in the face of the awful hardships that this war involved. What Belgium endured was bad enough, while the experiences of Poland, Servia, Armenia were an ascending scale of horrors, but also of humanity's power to stand suffering.

Life in the larger families of the olden times afforded more opportunities for the proper teaching of the place of suffering than in the smaller families of the modern time. Older children, as they grew up, had before them the example of mother's trials and hardships in bearing and rearing children, and so came to understand better the place of hard things in life. In a large family it was very rare when one or more of the members did not die, and thus growing youth was brought in contact with the greatest mystery in life, that of death. Very frequently at least one of the household and sometimes more, had to go through a period of severe suffering with which the others were brought in daily contact. It is sometimes thought in modern times that such intimacy with those who are suffering takes the joy out of life for those who are young, but any one who thinks so should consult a person who has had the actual experience; while occasionally it may be found that some one with a family history of this kind may think that he or she was rendered melancholy by it, nine out of ten or even more will frankly say that they feel sure that they were benefited. There is nothing in the world that broadens and deepens the significance of life like intimate contact with suffering, if not in person, then in those who are near and dear to us.

As a physician, I have often felt that I should like to take people who are constantly complaining of their little sorrows and trials, who are downhearted over some minor ailment, who sometimes suffer from fits of depression precipitated by nothing more, perhaps, than a dark day or a little humid weather, or possibly even a petty social disappointment, and put them in contact with cancer patients or others who are suffering severely day by day, yes, hour by hour, night and day, and yet who are joyful and often a source of joy to others. Let us not forget that nearly one hundred thousand people die every year from cancer in this country alone.

As a physician, I have often found that a chronic invalid in a house became the center of attraction for the whole household, and that particularly when it was a woman, whether mother or elder sister, all of the other members brought their troubles to her and went away feeling better for what she said to them. I have seen this not in a few exceptional instances, but so often as to know that it is a rule of life. Chronic invalids often radiate joy and happiness, while perfectly well people who suffer from minor ills of the body and mind are frequently a source of grumpiness, utterly lack sympathy, and are impossible as companions. An American woman, bedridden for over thirty years, has organized by correspondence one of the most beautiful charities of our time.

Pity properly restricted to practical helpfulness without any sentimentality is a beautiful thing. There is always a danger, however, of its arousing in its object that self-pity which is so eminently unlovely and which has so often the direct tendency to increase rather than decrease whatever painful conditions are present.

Crying over oneself is always to be considered at least hysterical. Crying, except over a severe loss, is almost unpardonable. It is often said that a good cry, like a rainstorm, clears the atmosphere of murk and the dark elements of life, but it is dangerous to have recourse to it. It is a sign of lack of character almost invariably and when indulged in to any extent will almost surely result in deterioration of the power to withstand the trials of life, whatever they may be.

Professor William James has suggested that not only should men and women stand the things that come to them in the natural course of events, but they should even go out and seek certain things hard to bear with the idea of increasing their power to withstand the unpleasant things of life. This is, of course, a very old idea in humanity, and the ascetics from the earliest days of Christianity taught the doctrine of self-inflicted suffering in order to increase the power of resistance.

It is usually said that the principal idea which the hermits and anchorites and the saintly personages of the early Middle Ages, of whose mortifications we have heard so much, had in inflicting pain on themselves was to secure merit for the hereafter. Something of that undoubtedly was in their minds, but their main purpose was quite literally ascetic. Ascesis, from the Greek, means in its strict etymology just exercise. They were exercising their power to stand trials and even sufferings, so that when these events came, as inevitably they would, seeing that we carry round with us what St. Paul called "this body of our death," they would be prepared for them.

Practically any psychologist of modern times who has given this subject any serious thought will recognize, as did Professor James, the genuine psychology of human nature that lies behind these ascetic practices. Nothing that I know is so thoroughgoing a remedy for self-pity as the actual seeking at times of painful things in order to train oneself to bear them. The old-fashioned use of disciplines, that is, little whips which were used so vigorously sometimes over the shoulders as to draw blood, or the wearing of chains which actually penetrated the skin and produced quite serious pain no longer seems absurd, once it is appreciated that this may be a means of bracing up character and making the real trials and hardships of life much easier than would otherwise be the case.

Not that human nature must not be expected to yield a little under severe trials and bend before the blasts of adverse fortune, but that there should not be that tendency to exaggerate one's personal feelings which has unfortunately become characteristic of at least the better-to-do classes in our time. Not that we would encourage stony grief, but that sorrow must be restrained and, above all, must not be so utterly selfish as to be forgetful of others.

Tears should, to a large extent, be reserved, as they are in most perfectly normal individuals, for joyous rather than sad occasions, for no one ever was supremely joyful without having tears in the eyes. It is when we feel most sympathetic to humanity that the gift of tears comes to us, and no feeling is quite so completely satisfying as comes from the tears of joy. Mothers who have heard of their boy's bravery, its recognition by those above him, and its reward by proper symbols, have had tears come welling to their eyes, while their hearts were stirred so deeply with sensations of joy and pride that probably they have never before felt quite so happy.

CHAPTER VI

AVOIDANCE OF CONSCIOUS USE OF THE WILL

"Our bodies are our gardens to which our wills are gardeners."
Othello

Doctor Austin O'Malley, in his little volume, "Keystones of Thought", says: "When you are conscious of your stomach or your will you are ill." We all appreciate thoroughly, as the result of modern progress in the knowledge of the influences of the mind on the body, how true is the first part of this saying, but comparatively few people realize the truth of the second part. The latter portion of this maxim is most important for our consideration. It should always be in the minds of those who want to use their own wills either for the purpose of making themselves well, or keeping themselves healthy, but above all, should never be forgotten by those who want to help others get over various ills that are manifestly due in whole or in part to the failure to use the vital energies in the body as they should be employed.

Conscious use of the will, except at the beginning of a series of activities, is always a mistake. It is extremely wasteful of internal energy. It adds greatly to the difficulty of accomplishing whatever is undertaken. It includes, above all, watching ourselves do things, constantly calculating how much we are accomplishing and whether we are doing all that we should be doing, and thus makes useless demands on power partly by diversion of attention, partly by impairment of concentration, but above all by adding to the friction because of the inspection that is at work.

The old kitchen saw is "a watched kettle never boils." The real significance of the expression is of course that it seems to take so long for the water to boil that we become impatient while watching and it looks to us as though the boiling process would really never occur. This is still more strikingly evident when we are engaged in watching our own activities and wondering whether they are as efficient as they should be. The lengthening of time under these circumstances is an extremely important factor in bringing about tiredness. Ask any human being unaccustomed to note the passage of time to tell you when two minutes have elapsed; inevitably he will suggest at the end of thirty to forty seconds that the two minutes must be up. Only by counting his pulse or by going through some regular mechanical process will he be enabled to appreciate the passage of time in anything like its proper course. When watched thus, time seems to pass ever so much more slowly than it would otherwise.

It is extremely important then that people should not acquire the impression that they must be consciously using their will to bring themselves into good health and keep themselves there, for that will surely defeat their purpose. What is needed is a training of the will to do things by a succession of harder and harder tasks until the ordinary acts of life seem comparatively easy. Intellectual persuasion as to the efficiency of the will in this matter means very little. The ordinary feeling that reasoning means much in such matters is a fallacy. Much thinking about them is only disturbing of action as a rule and Hamlet's expression that the "native hue of resolution is sicklied o'er with the pale cast of thought" is a striking bit of psychology.

Shakespeare had no illusions with regard to the place of the will in life and more than any English author has emphasized it. I have ventured to illustrate this by quotations from him under each chapter heading, but there are many more quite as applicable that might readily be found. He knew above all how easy it was for human beings to lessen the power of their wills and has told us of "the cloy'd will that satiateth unsatisfied desire" and "the bridles of our wills", and has given us such adjectives as "benumbed" and "neutral" and "doting", which demonstrates his recognition of how men weaken their wills by over-deliberation.

The mode of training in the army is of course founded on this mode of thinking. The young men in the United States Army want to accomplish every iota of their duty and are not only willing but anxious to do everything that is expected of them. There were some mighty difficult tasks ahead of them over in Europe and our method of preparing the men was not by emphasizing their duty and dinning into their ears and minds how great the difficulty would be and how they must nerve themselves for the task. Such a mode of preparation would probably have been discouraging rather than helpful. But they were trained in exercises of various kinds in an absolutely regular life under plain living in the midst of hard work until their wills responded to the word of command quite unconsciously and immediately without any need of further prompting. Their bodies were trained until every available source of energy was at command, so that when they wanted to do things they set about them without more ado, and as they were used to being fatigued they were not constantly engaged in dreading lest they should hurt themselves, or fostering fears that they might exhaust their energies or that their tiredness, even when apparently excessive, would mean anything more than a passing state that rest would repair completely.

If at every emergency of their life at war soldiers had to go through a series of conscious persuasions to wake up their will and set their energies at work, and if they had to occupy themselves every time in presenting motives why this activity should not be delayed, then military discipline, at least in so far as it involves prompt obedience, would almost inevitably be considerable of a joke. What is needed is unthinking, immediate obedience, and this can be secured only by the formation of deeply graven habits which enable a man to set about the next thing that duty calls for at once.

Every action that we perform is the result of an act of the will, but we do not have to advert to that as a rule; whenever any one gets into a state of mind where it is necessary to be constantly adverting to it, then, as was said at the beginning of this chapter, there is something the matter with the will. The faculty is being hampered in its action by consciousness, and such hampering leads to a great waste of energy.

The will is the great, unconscious faculty in us. By far the greater part of what has come unfortunately to be called the unconscious and the subconscious and that has occupied so much of the attention of modern writers on psychological subjects is really the will at work. It attains its results we know not how, and it is prompted to their accomplishment in ways that are often very difficult for us to understand. Its effects are often spoken of as due to the submerged self or the subliminal self or the other self, but it is only in rare and pathological cases as a rule that such expressions are justified once the place of the will is properly recognized.

It is often said, for instance, that the power some people have of waking after a certain period of sleep at night or after a short nap during the daytime, a power that a great many more people would possess if only they deliberately practised it, is due to the subconsciousness or the subliminal personality of the individual which wakes him up at the determined time. Why those terms should be used when other things are accomplished by the human will just as mysteriously is rather difficult to understand. It is well recognized that if an individual in the ordinary waking state wants to do something after the lapse of an hour or so he will do it, provided his will is really awake to the necessity of accomplishing it. It is true that he may become so absorbed in his current occupation as to miss the time, but such abstraction usually means that he was not sufficiently interested in the duty that was to be performed as to keep the engagement with himself, or else that he is an individual in whom the intellectual overshadows the voluntary life. We speak of him as an impractical man.

We all know the danger there is in putting off calling some one by telephone on being told that "the line is busy", for not infrequently it will happen that several hours will elapse before we think of the matter again and then perhaps it may be too late. If we set a definite time limit with ourselves, however, then our will will prompt us quite as effectively, though quite as inexplicably, at the expiration of that time as it awakes those who have resolved to be aroused at a predetermined moment. We may miss our telephone engagement with ourselves, but we practically never miss an important train, because having deeply impressed upon ourselves the necessity for not missing this, our will arouses us to activity in good time. There is not the slightest necessity, however, for appealing to the unconscious or the subconscious in this. It is true that there is a wonderful sentinel within us that awakes us from daydreams or disturbs the ordinary course of some occupation to turn our attention to the next important duty that we should perform. We know that this sentinel is quite apart from our consciousness; but the power we have of setting ourselves to doing anything is exemplified in very much the same way. When I want a book, I do not know what it is that sets my muscles in motion and brings me to a shelf and then directs my attention to choosing the one I shall take down and consult. It is an unconscious activity, but not the activity of unconsciousness, which is only a contradiction in terms. [Footnote]

[Footnote : It is true that there is a particular phase of our intellectual effort included under the modern terms unconscious or subconscious that is mysterious enough to deserve a special name, but we already have an excellent term for this quality which is not vague but thoroughly descriptive of its activity. This is intuition, a word that has been in use for nearly a thousand years now and signifies the immediate perception of a truth, by a flash as it were. We may know nothing about a subject and may have only begun to think about it, when there flashes on us a truth that has perhaps never occurred to any one else and certainly has never been in our minds before. It has been suggested in recent years that such flashes of intelligence are due to the secondary personality or the subliminal self or the other self, and it is often added that it is the development of our knowledge of these phases of psychology that represents modern progress in the science of mind. Only the term for it is new, however, for intuition has been the subject of special intensive study for a long while. Indeed, the reason why the old-time poet appealed to the muses for aid and the modern poet suggests inspiration as the source of his poetic thought, is because both of them knew that their best thoughts flash on them, not as the result of long and hard thinking, but by some process in which with the greatest facility come perceptions that even they themselves are surprised to learn that they have. To say that such things come from the unconscious is simply to ignore this wonderful power of original thought, that is, primary perception. Emerson suggested that intuition represented all the knowledge that came without tuition, as if this were the etymology, and the hint is excellent for the meaning, though the real derivation of the word has no relation to tuition. To attribute these original thoughts to the unconscious or any partly conscious faculty in us is to ignore a great deal of careful study of psychology before our time. It is besides to entangle oneself in the absurdity of discussing an unconscious consciousness.]

While many people are inclined to feel almost helpless in the presence of the idea that it is their unconscious selves that enable them to do things or initiate modes of activity, the feeling is quite different when we substitute for that the word "will." All of us recognize that our wills can be trained to do things, and while at first it may require a conscious effort, we can by the formation of habits not only make them easy, but often delightful and sometimes quite indispensable to our sense of well being. Walking is extremely difficult at the beginning, when its movements are consciously performed, but it becomes a very satisfying sort of exercise after a while and then almost literally a facile, nearly indispensable activity of daily life, so that we feel the need for it, if we are deprived of it.

This has to be done with regard to the activities that make for health. We have to form habits that render them easy, pleasant, and even necessary for our good feelings. This can be done, as has been suggested in the chapter on habits, but we have to avoid any such habit as that of consciously using the will. That is a bad habit that some people let themselves drop into but it should be corrected. Having set our activities to work we must, as far as possible, forget about them and let them go on for themselves. It is not only possible but even easy and above all almost necessary that we should do this. Hence at the beginning people must not expect that they will find the use of their will easy in suppressing pain, lessening tiredness, and facilitating accomplishment, but they must look forward to the time quite confidently when it will be so. In the meantime the less attention paid to the process of training, the better and more easily will the needed habits be formed.

Failure to secure results is almost inevitable when conscious use of the will comes into the problem. As a rule a direct appeal should not be made to people to use their wills, but they should be aroused and stimulated in various ways and particularly by the force of example. What has made it so comparatively easy for our young soldiers to use their wills and train their bodies and get into a condition where they are capable of accomplishing what they would have thought quite impossible before, has been above all the influence of example. A lot of other young men of their own age are standing these things exemplarily. They are seen performing what is expected of them without complaint, or at least without refusal, and so every effort is put forth to do likewise without any time spent on reflection as to how difficult it all is or how hard to bear or how much they are to be pitied. It is not long before what was hard at first becomes under repetition even easy and a source of fine satisfaction. Getting up at five in the morning and working for sixteen hours with only comparatively brief intervals for relaxation now and then, and often being burdened with additional duties of various kinds which must be worked in somehow or other, seems a very difficult matter until one has done it for a while. Then one finds everything gets done almost without conscious effort. Will power flows through the body and lends hitherto unexpected energy, but of this there is no consciousness; indeed, conscious reflection on it would hamper action. No wonder that as a result of the facility acquired, one comes to readily credit the assumption that the will is a spiritual power and that some source of energy apart from the material is supplying the initiative and the resources of vitality that have made accomplishment so much easier than would have been imagined beforehand. This is quite literally what training of the will means: training ourselves to use all our powers to the best advantage, not putting obstacles in their way nor brakes on their exertion, but also not thinking very much about them or making resolutions. The way to do things is to do them, not think about them.

Professor James is, as always, particularly happy in his mode of expressing this great truth. He insists that the way to keep the will active is not by constantly thinking about it and supplying new motives and furbishing up old motives for its activity, but by cultivating the faculty of effort. His paragraph in this regard is of course well known, and yet it deserves to be repeated here because it represents the essence of what is needed to make the will ready to do its best work. He says:

"As a fine practical maxim, relative to these habits of the will, we may, then, offer something like this: Keep the faculty of effort alive in you by a little gratuitous exercise every day. That is, be systematically ascetic or heroic in little unnecessary points, do every day or two something for no other reason than that you would rather not do it, so that when the hour of dire need draws nigh, it may find you not unnerved and untrained to stand the test. Asceticism of this sort is like the insurance which a man pays on his house and goods. The tax does him no good at the time and possibly may never bring him a return. But if the fire does come, his having paid it will be his salvation from ruin. So with the man who has daily inured himself to habits of concentrated attention, energetic volition, and self-denial in unnecessary things. He will stand like a tower when everything rocks around him, and when his softer fellow mortals are winnowed like chaff in the blast."

To do things on one's will without very special interest is an extremely difficult matter. It can be done more readily when one is young and when certain secondary aims are in view besides the mere training of the will, but to do things merely for will training becomes so hard eventually that some excuse is found and the task is almost inevitably given up. Exercising for instance in a gymnasium just for the sake of taking exercise or keeping in condition becomes so deadly dull after a while that unless there is a trainer to keep a man up to the mark, his exercise dwindles from day to day until it amounts to very little. Men who are growing stout about middle life will take up the practice of a cold bath after ten minutes or more of morning exercises with a good deal of enthusiasm, but they will not keep it up long, or if they do continue for several months, any change in the daily routine will provide an excuse to drop it. Companionship and above all competition in any way greatly helps, but it takes too much energy of the will to make the effort alone. Besides, when the novelty has worn off and routine has replaced whatever interest existed in the beginning in watching the effect of exercise on the muscles, the lack of interest makes the exercise of much less value than before. If there is not a glow of satisfaction with it, the circulation, especially to the periphery to the body, is not properly stimulated and some of the best effect of the exercise is lost.

Athletes often say of solitary exercise that it leaves them cold, which is quite a literal description of the effect produced on them. The circulation of the surface is not stimulated as it is when there is interest in what is being done and so the same warmth is not produced at the surface of the body.

It is comparatively easy to persuade men who need outdoor exercise to walk home from their offices in the afternoon when the distance is not too far, but it is difficult to get them to keep it up. The walk becomes so monotonous a routine after a time that all sorts of excuses serve to interrupt the habit, and then it is not long before it is done so irregularly as to lose most of its value. Here as in all exercise, companionship which removes conscious attention from advertence to the will greatly aids. On the other hand, as has been so clearly demonstrated in recent years, it is very easy to induce men to go out and follow a little ball over the hills in the country, an ideal form of exercise, merely because they are interested in their score or in beating an opponent. Any kind of a game that involves competition makes people easily capable of taking all sorts of trouble. Instead of being tired by their occupation in this way and not wanting to repeat it, they become more and more interested and spend more and more time at it. The difference between gymnastics and sport in this regard is very marked.

In sport the extraneous interest adds to the value of the exercise and makes it ever so much easier to continue; when it sets every nerve tingling with the excitement of the game, it is doing all the more good. Gymnastics grow harder unless in some way associated with competition, or with the effort to outdo oneself, while indulgence in sport becomes ever easier. Many a young man would find it an intolerable bore and an increasingly difficult task if asked to give as much time and energy to some form of hard work as he does to some sport. He feels tired after sport, but not exhausted, and becomes gradually able to stand more and more before he need give up, thus showing that he is constantly increasing his muscular capacity.

Conscious training of the will is then practically always a mistake. It is an extremely difficult thing to do, and the amount of inhibition which accumulates to oppose it serves after a time to neutralize the benefit to be derived. Good habits should be formed, but not merely for the sake of forming them. There should be some ulterior purpose and if possible some motive that lifts men up to the performance of duty, no matter how difficult it is.

Our young men who went to the camps demonstrated how much can be accomplished in this manner. They were asked to get up early in the morning, to work hard for many hours in the day, or take long walks, sometimes carrying heavy burdens, and were so occupied that they had but very little time to themselves. They were encouraged to take frequent cold baths, which implied further waste of heat energy, and then were very plainly fed, though of course with a good, rounded diet, well-balanced, but without any frills and with very little in it that would tempt any appetite except that of a hungry man. They learned the precious lesson that hunger is the best sauce for food.

Most of these men were pushed so hard that only an army officer perfectly confident of what he was doing and well aware that all of his men had been thoroughly examined by a physician and had nothing organically wrong with them would have dared to do it. A good many of us had the chance to see how university men took the military regime. Long hours of drilling and of hard work in the open made them so tired that in the late afternoon they could just lie down anywhere and go to sleep. I have seen young fellows asleep on porches or in the late spring on the grass and once saw a number of them who found excellent protection from the sun in what to them seemed nice soft beds at least they slept well in them inside a series of large earthen-ware pipes that were about to be put down for a sewer. Some of them were pushed so hard, considering how little physical exercise they had taken before, that they fainted while on drill. Quite a few of them were in such a state of nervous tension that they fainted on being vaccinated. Almost needless to say, had they been at home, any such effect would have been a signal for the prompt cessation of such work as they were doing, for the home people would have been quite sure that serious injury would be done to their boys. These young fellows themselves did not think so. Their physicians were confident that with no organic lesion present the faint was a neurotic derangement and not at all a symptom of exhaustion. The young soldiers would have felt ashamed if there had been any question of their stopping training. They felt that they could make good as well as their fellows. They would have resented sympathy and much more pity. They went on with their work because they were devoted to a great cause. After a time, it became comparatively easy for them to accomplish things that would hitherto have been quite impossible and for which they themselves had no idea that they possessed the energy. It was this high purpose that inspired them to let more and more of their internal energy loose without putting a brake on, until finally the habit of living up to this new maximum of accomplishment became second nature and therefore natural and easy of accomplishment.

Here is the defect in systems which promise to help people to train their wills by talking much about it, and by persuading them that it can be done, that all they have to do is to set about it. Unless one has some fine satisfying purpose in doing things, their doing is difficult and fails to accomplish as much good for the doer as would otherwise be the case. Conscious will activity requires, to use old-fashioned psychological terms, the exercise of two faculties at the same time, the consciousness and the will. This adds to the difficulty of willing. What is needed is a bait of interest held up before the will, constantly tempting it to further effort but without any continuing consciousness on the part of the individual that he must will it and keep on willing it. That must ever be a hampering factor in the case. Human nature does not like imperatives and writhes and wastes energy under them. On the contrary, optatives are pleasant and give encouragement without producing a contrary reaction; and it is this state of mind and will that is by far the best for the individual.

Above all, it is important that the person forming new habits should feel that there is nothing else to be done except the hard things that have been outlined. If there is any mode of escape from the fulfillment of hard tasks, human nature will surely find it. If our young soldiers had felt that they did not have to perform their military duties and that there was some way to avoid them, the taking of the training would have proved extremely difficult. They just had to take it; there was no way out, so they pushed themselves through the difficulties and then after a time they found that they were tapping unsuspected sources of energy in themselves. For when people have to do things, they find that they can do ever so much more than they thought they could, and in the doing, instead of exhausting themselves, they actually find it easier to accomplish more and more with ever less difficulty. The will must by habit be made so prompt to obey that obedience will anticipate thought in the matter and sometimes contravene what reason would dictate if it had a chance to act. The humorous story of the soldier who, carrying his dinner on a plate preparatory to eating it, was greeted by a wag with the word "Attention!" in martial tones, and dropped his dinner to assume the accustomed attitude, is well known. Similar practical jokes are said to have been played, on a certain number of occasions in this war, with the thoroughly trained young soldier.

The help of the will to the highest degree is obtained not by a series of resolutions but by doing whatever one wishes to do a number of times until it becomes easy and the effort to accomplish it is quite unconscious. Reason does not help conduct much, but a trained will is of the greatest possible service. It can only be secured, however, by will action. The will is very like the muscles. There is little use in showing people how to accomplish muscle feats; they must do them for themselves. The less consciousness there is involved in this, the better.

CHAPTER VII

WHAT THE WILL CAN DO

"I can with ease translate it to my will."
King John

It should be well understood from the beginning just what the will can do in the matter of the cure or, to use a much better word, the relief of disease, not forgetting that disease means etymologically and also literally discomfort rather than anything else. The will cannot cure organic disease in the ordinary sense of that term. It is just as absurd to say that the will can bring about the cure of Bright's disease as it is to suggest that one can by will power replace a finger that has been lost. When definite changes have taken place in tissues, above all when connective tissue cells have by inflammatory processes come to take the place of organic tissue cells, then it is idle to talk of bringing about a cure, though sometimes relief of symptoms may be secured; above all the compensatory powers of the body may be called upon and will often bring relief, for a time, at least. What is true of kidney changes applies also to corresponding changes in other organs, and there can be no question of any amount of will power bringing about the redintegration of organs that have been seriously damaged by disease or replacing cells that have been destroyed.

There are however a great many organic diseases in which the will may serve an extremely useful purpose in the relief of symptoms and sometimes in producing such a release of vital energy previously hampered by discouragement as will enable the patient to react properly against the disease. This is typically exemplified in tuberculosis of the lungs. Nothing is so important in this disease, as we shall see, as the patient's attitude of mind and his will to get well. Without that there is very little hope. With that strongly aroused, all sorts of remedies, many of them even harmful in themselves, have enabled patients to get better merely because the taking of them adds suggestion after suggestion of assurance of cure. The cells of the lungs that have been destroyed by the disease are not reborn, much less recreated, but nature walls off the diseased parts, and the rest of the lungs learn to do their work in spite of the hampering effect of the diseased tissues. When fresh air and good food are readily available for the patient, then the will power is the one other thing absolutely necessary to bring about not only relief from symptoms, but such a betterment in the tissues as will prevent further development of the disease and enable the lungs to do their work. The disease is not cured, but, as physicians say, it is arrested, and the patient may and often does live for many years to do extremely useful work.

In a disease like pneumonia the will to get well, coupled with the confidence that should accompany this, will do more than anything else to carry the patient over the critical stage of the affection. Discouragement, which is after all by etymology only disheartenment, represents a serious effect upon the heart through depression. The fullest power of the heart is needed in pneumonia and discouragement puts a brake on it. As we shall see it is probably because whiskey took off this brake and lifted the scare that it acquired a reputation as a remedy in pneumonia and also in tuberculosis. In spite of what was probably an unfavorable physical effect, whiskey actually benefited the patient by its production of a sense of well being and absence of regard for consequences. Hence its former reputation. This extended also to its use in a continued fever where the same disheartenment was likely to occur with unfortunate consequences on the general condition and above all with disturbance of appetite and of sleep. Worry often made the patients much more restless than they would otherwise have been and they thus wasted vital energy needed to bring about the cure of the affection under which they were laboring.

In all of these cases solicitude led to surveillance of processes within the body and interfered with their proper performance. It is perfectly possible to hamper the lungs by watching their action, and the same thing may be done for the heart. Whenever involuntary activities in the body are watched, their proper functioning is almost sure to be disturbed. We have emphasized that in the chapter on "Avoidance of Conscious Use of the Will," and so it need not be dwelt on further here.

Even apart from over-consciousness there occur some natural dreads that may disturb nature's vital reactions, and these can be overcome through the will. There is a whole series of inhibitions consequent upon fears of various kinds that sadly interfere with nature's reaction against disease. To secure the neutralization of these the will must be brought into action, and this is probably better secured by suggestion, that is, by placing some special motive before the individual, than by any direct appeal. Particularly is this true if patients have not been accustomed before this to use their wills strenuously, for they will probably be disturbed by such an appeal.

What will power when properly released can do above all is to bring the relief of discomfort. In a great many cases the greater part of the discomfort is due to over-sensitization and over-attention. Even in such severe organic diseases as cancer, the awakening of the will may accomplish very much to bring decided relief. This is why we have had so many "cancer cures" that have failed. They made the patient feel better at first, and they relieved pain to some extent and therefore were thought to be direct remedial agents for the cancer itself. The malignant condition however has progressed without remission, though sometimes, possibly as the result of the new courage given flowing as surplus vitality into the tissues, perhaps the progress of the lesion has been retarded. The patient sometimes has felt so much better as to proclaim himself cured. What is thus true of cancer will be found to occur in any very serious organic condition, such as severe injury, chronic disease involving important organs, and even such nutritional diseases as anemia or diabetes. The awakening in the patient of the feeling that there is hope and the maintenance of that hope in any way will always bring relief and usually some considerable remission in the disease.

It is in convalescence above all, however, that the will power manifests its greatest helpfulness. When patients are hopeful and anxious to get well they are tempted to eat properly, to get out into the air; they thus sleep better and recovery is rapid. Whenever they are disheartened, as for instance when husband and wife have been together in an injury, or both have contracted a disease and one of them dies, the survivor is likely to have a slow and lingering convalescence. The reason is obvious: there is literal lack of will power or at least unwillingness to face the new conditions of life, and vitality is spent in vain regret for the companionship that has been lost. This depression can only be lifted by motives that appeal to the inner self and by such an awakening of the will for further interests in life as will set vital energies flowing freely again.

In convalescence from injuries received after middle life or from affections that have been accompanied by incapacity to use muscles there is particular need of the will. A great many older people refuse to go through the pain and discomfort, soreness and tenderness as the younger folk who are training their muscles call them, which must be borne in order to bring about redevelopment of muscles, after they have once become atrophic from disuse. The refusal to push through a period of what is often rather serious discomfort leads many people to foster disabilities and use their muscles in wrong ways sometimes even for years. Something occurs then to arouse their wills and they get better. Anything that will do this will cure them. Sometimes it is a new liniment, sometimes a new mode of manipulation or massage, sometimes some supposed electrical or magnetic discovery and sometimes the touch of a presumed healer. Anything at all will be effective provided it wakens their wills into such activity as will enable them to persist in the use of their muscles through the period of soreness and tenderness necessary to restore proper muscular functions.

It is quite surprising to see what can be accomplished in this way, and the quacks and charlatans of the world have made their fortunes out of such patients always, while their cure has been the greatest possible advertisement and has attracted ever so many other patients to these so-called healers. Nothing that can be done for these patients will have any good results unless their own wills are aroused, new hope given them and they themselves made to tap the layers of energy in them that can restore them to health. To tell them that they were to be cured by their own will, however, would probably inhibit utterly this energy that is needed, so that somehow they have to be brought to the state of mind in which they will accomplish the purpose demanded of them by indirection.

The will is particularly capable of removing obstacles to nutrition that have often hampered the activities and sometimes seriously impaired the health of patients. Many people are not eating enough for one reason or another and need to have their diet regulated, not in the direction of a limitation or selection of food, though this appeals to so many people under the term dieting, but so that they shall eat enough and of the proper variety to maintain their health and bodily functions. A great many nervous diseases are dependent on lack of sufficient food. Eating in those who lead sedentary lives much indoors is ever so much more a matter of will than of appetite. When people say that they eat all they want to, what they mean, as a rule, is that they eat all that they have formed the habit of eating. Other habits can readily be formed and will often do them good. For a great many of the less serious symptoms which make people valetudinarians, nervous indigestion, insomnia, tendencies to headache, queer feelings in the head, constipation, the proper habit secured by will power, of eating so as to secure sufficient food, is the most important single factor. This the will must be trained to accomplish.

Now that disease prevention has become even more important than cure, the will is an extremely efficient element. Air, food, exercise are important factors for healthy living. A great many people are neglecting them and then seem surprised that they should suffer from various symptoms of impaired functioning of bodily organs. Many men and a still greater number of women are staying in the house so much that their oxidation within the body is at a low ebb, and it is no wonder that vital processes are not carried on to the best advantage. Our generation has eliminated exercise from life to a great extent, and now that the auto and the trolley car limit walking, not only the feet of mankind suffer severely, but all the organs in the body work at a disadvantage for lack of the exercise that they should have. No wonder that under the circumstances appetite is impaired and other functions of the body suffer. Instead of simple foods various artificial stimulants are employed such as alcohol, spices, and the like to provoke appetite, often with serious consequences for the digestive organs. The will to be well includes the willing of the means proper to that purpose, and particularly regular exercise, several hours a day in the air, good simple food taken in sufficient quantity at three regular intervals and the avoidance of such sources of worry as will disturb physical functions.

CHAPTER VIII

PAIN AND THE WILL

"That the will is infinite and the execution confined."
Troilus and Cressida

The symptom of disease that humanity dreads the most is pain. Fortunately, it is also the symptom which is most under the control of the will, and which can be greatly relieved by being bravely faced and, to as great an extent as possible, ignored. It requires courage and usually persistent training to succeed in the relief of severe pain in this way, but men have done it, and women too, and men and women can do it, if they really want to, though unfortunately all of the trend of modern life has been in the opposite direction, of avoiding pain at whatever cost instead of bravely facing it. The American Indian, trained from his youth to stand severe pain, scoffed at even the almost ingeniously diabolical tortures of his enemy captors. After they had pushed slivers beneath his nails or slowly crushed the end of a finger, or put salt in long, superficial wounds that had bared a whole series of sensitive cutaneous nerves, he has been known to laugh at them, and ask them proudly, without giving a sign of the pain that he was enduring, whether that was all that they could do. It was just a question of the human will overcoming even the worst sensations that the body could send up to the brain and deliberately refusing to permit any reactions that would reveal the reflex torment that was actually taking place.

The war has done much to bring back the recognition of that diminution to a great extent at least or even almost entire suppression of pain which may occur, indeed almost constantly does occur, as a consequence of a man facing it bravely. We have been accustomed to think of the early martyrs as probably divinely helped in their power to withstand pain. Whatever of celestial aid they had, we know that martyrs for all sorts of causes, some of them certainly not divine, have exhibited some degree of this same steadfastness. Their behavior makes it reasonably clear that as the result of making up their minds to stand the pain involved, they have actually suffered so little that it was not difficult to suppress external manifestations of their sufferings. It is not merely a suppression of the reflexes that has occurred but a minimizing to a very striking degree of the actual sensations felt. We have many stories of the older time before the modern use of anaesthetics, which tell how bravely men endured pain and at the same time retained their power to do things. Indeed, some of them accomplished purposes in the midst of what would seem like supreme agony which made it very clear that pain alone has nothing like the prostrating effect that it is often supposed to have.

For we have well authenticated tales of physicians performing amputations on themselves at times when no other assistance was available, and accomplishing the task so well that they recovered without complications. A blacksmith in the distant West, whose leg had been crushed by the fall of a huge beam, actually had himself carried into his shop and amputated his own limb above the knee, searing the blood vessels with hot irons as he proceeded. Such a manifestation of will power is, of course, exceptional to a degree, and yet it illustrates what men can do in the face of conditions that are usually supposed to be overwhelming. Many a man in lumber camps or in distant island fisheries or on board fishing vessels, far beyond the hope of reaching a physician in time for him to be of service, has done things of this kind. We can be quite sure that the will to accomplish for himself what seemed necessary to save his life lessened his pain, made it ever so much more bearable and generally proved the power of the human will over even these physical manifestations in the body that are commonly supposed to be quite beyond any interference from the psychical part of nature. The spirit can still dominate the flesh, even in matters of pain, and dictate how much it shall be affected. It is a hard lesson to learn, but it is one that can be learned by proper persistence.

In the early part of the war particularly many a young man had to face even serious operations without an anaesthetic. The awful carnage of the first six weeks of the war had not been anticipated and therefore there were not sufficient stores of anaesthetics available to permit of their use in every case. Besides, many operations had to be performed so close to the front and under such circumstances that there could not be anaesthetics for all of them; and it was a never-ending source of surprise to those who witnessed the details to see how bravely and uncomplainingly the young men took their enforced suffering. Many a one, when his turn came to be operated on, quietly asked for a cigarette and then bore unflinchingly painful manipulations that the surgeon was extremely sorry to have to inflict. Over and over again, when there was question of the regular succession of patients, young soldiers in severe pain suggested that some one else who seemed in worse condition than they, or who perhaps was not quite so well able to stand pain and control himself, should be attended to before they were. There is no doubt at all that this very power of self-control lessened their pain and made it ever so much easier to bear and less of a torment than it would have been otherwise.

Any great diversion of mind that turns the attention completely to something else will lessen even severe pain so much as to make it quite negligible for the moment. Headaches disappear promptly when there is an alarm of fire, and toothaches have been known to vanish, for the time at least, as the result of a burglar scare. Much less than this is needed, however, and there are many familiar examples which illustrate the fact that the turning of the attention to something else will greatly diminish or even abolish pain.

The well known story of the French surgeon about to set a dislocation is a typical demonstration. His patient was a woman of the nobility, her dislocation was of the shoulder and it was necessary for him to inflict very severe pain in order to replace it. Besides, as the result of the reflex of that pain, he was certain to meet with great resistance from spasm in the surrounding muscles. It was before the days of anaesthetics, which relieve all of these inconveniences, and above all, relax the muscles. The surgeon got ready to do the ultimate manipulation that would replace the joint in its proper relation, and necessarily inflicted no little pain in his preparations. The lady complained very much, so he turned on her angrily, told her that she must stand it, slapped her in the face, and before she had recovered from the shock, the dislocation had been restored to the normal condition. It was rather heroic treatment, and it is to be hoped that she understood it, but it is easy to understand how much the procedure lessened her physical pain.

When the mind is very much preoccupied and the will intent on accomplishing some immediate purpose, even severe pain will not be felt at all. Instances of this are not rare, and men who are advancing in a charge on a battlefield will often be wounded rather severely, and yet continue to advance without knowing anything about their wounds until a friend calls attention to their bleeding, or they themselves notice it; or perhaps even loss of blood may make them faint. The late President Roosevelt furnished a magnificent illustration of this principle when he was wounded some years ago in the midst of a political campaign. A crank shot at him, in one of the Western cities, and though the bullet penetrated four inches of muscle on his chest wall, and then flattened itself against a rib, he did not know that he was wounded. The flattening of the bullet must have represented at least as much force as would be exerted by a heavy blow on the chest, and yet the Colonel never felt it. His friends congratulated him on his escape from injury until it was noted that blood was oozing through a hole that had been made in his coat. The intense will activity of the President simply kept him from noticing either the shock or the pain.

Not long before the war a striking example was given of how a man may stand suffering in spite of long years of the refining influences of a sedentary scholarly life, most of it spent indoors. The second last General of the Jesuits developed a sarcoma on his upper arm and was advised to submit to an amputation of the arm at the shoulder joint. He was a man well on in the sixties and the operation presented an extremely serious problem. The surgeons suggested that he should be ready for the anaesthetic at a given hour the next morning and then they would proceed to operate. He replied that he would be ready for the operation at the time suggested, but that he would not take an anaesthetic. They argued with him that it would be quite impossible for him to stand unanaesthetized the extensive cutting and dissection necessary to complete an operation of this kind in an extremely important part of the body, where large nerves and arteries would have to be cut through and where the slightest disturbance on the part of the patient might easily lead to serious or even fatal results. Above all, he could not hope to stand it in tissues that had been rendered more sensitive than before by the enlarged circulation to the part, due to the growth of the tumor, and the consequent hyperaemic condition of most of the tissues through which the cutting would have to be done and which were thus hypersensitized.

He insisted, however, that he would not take an anaesthetic, for surely here seemed a chance to welcome suffering voluntarily as his Lord and Master had done. I believe that the head surgeon said at first that he would not operate. He felt sure that the operation would have to be interrupted after it had been begun, because the patient would not be able to stand the pain and there would then be the danger from bleeding as well as from infection which might occur. The General of the Jesuits, however, was so calm and firm that at last it was determined to permit him to try at least to stand it, though most of the surgeons were sure that he would probably have to give up and allow himself to be anaesthetized before they were through.

The event then was most interesting. The patient not only underwent the operation without a murmur, but absolutely without wincing. The surgeon who performed the operation said afterwards, "It was like cutting wax and not human flesh, so far as any reaction was concerned, though of course it bled."

The story carries its lesson of the power of a brave man to face even such awful pain as this and probably actually overcome it to such an extent that he scarcely felt it, simply because he willed that he would do so and occupied himself with other thoughts during the process.

Such an example as that of this General of the Jesuits will seem to most people a reversion to that mystical attitude of mind of the medieval period, when somehow or other people were able to stand ever so much more pain than any one in our time could possibly think of enduring. We hear of saints of the Middle Ages who inflicted what now seem hideous self-tortures on themselves and not only bore them bravely but went about life smiling and doing good to others while they were under the influence of them. It would seem quite impossible, however, for people of the modern time to get into any such state of mind. Our discoveries for the prevention of pain have made it unnecessary to stand much suffering, and as a result mankind would seem to have lost some if not most of the faculty of standing pain. So little of truth is there in any such thought that any number of the young men of the present generation between twenty and thirty, that is, during the very years when mankind most resents pain and therefore reacts most to it, and by the same token feels it the most, have shown during this war that they possessed all the old-fashioned faculty of standing pain without a whimper and thinking of others while they did it.

Lack of advertence always lessens pain and may even nullify it until it becomes exceedingly severe. In his little volume, "A Journey around My Room", Xavier de Maistre dwells particularly on the fact that his body, when his spirit was wandering, would occasionally pick up the fire tongs and burn itself before his alter ego could rescue it. Concentration of attention on some subject that attracts may neutralize pain and make it utterly unnoticed until physical consequences develop. Undoubtedly dwelling on pain, anticipating it, noting the first sensations that occur, multiplies the painful feeling. The physical reasons for this are to be found in the increased blood supply consequent upon conscious attention to any part, which sensitizes the nerves of the area and the added number of nerve fibers that are at once put into association with the area by the act of concentration of the attention. These serve to render sensation much more acute than it would otherwise be. It might seem impossible to control the attention, but this has been done over and over again, even in the midst of severe pain, until there is no doubt that it is quite possible. As for the increase of pain by deliberate attention, that is so familiar an experience that practically every one has had it at some time.

The reason for it has become very clear as the result of our generation's investigations into the constitution of the nervous system. The central nervous system, instead of being a continuum, or series of nerve elements which are directly connected with each other, consists of a very large number of separate individual cells which only make contacts with each other, the nerve impulses flowing over across the contact. The demonstration of these we owe originally to Ramon y Cajal, the distinguished Spanish brain anatomist, to whom was awarded some years ago the Nobel Prize as well as the Prize of the City of Paris for his researches.

In connection with his surprising discoveries as to the neurons which make up the brain, he suggested the Law of Avalanche, which would serve to explain the supersensitiveness of parts to which concentrated attention is paid. According to this law, pain felt in any small area of the body may be multiplied very greatly if the sensation from it is distributed over a considerable part of the brain, as happens when attention is centered upon it. A pain message that comes from a localized area of the body disturbs under normal conditions at most a few thousand cells in the brain, because the area is directly represented only by these cells. They are connected however by dendrites and cell branches of various kinds with a great many other cells in different parts of the brain. A pain message that comes up will ordinarily produce only disturbance of the directly connected cells, but it may be transmitted and diffused over a great many of the cells of the cortex of the brain if the attention is focused strongly on it. The area at first affected, but a few thousand cells, may spread to many millions or perhaps even some hundreds of millions of them, if the centering of attention causes them to be "connected up", as the electricians say, with the originally affected small group of cells.

It is just what happens in high mountains when a few stones loosened somewhere near the top by the wind or by melting processes begin their course down the mountain side. On the way they disturb ever more and more of the loose pieces of ice and the shifting snows as well as the rocks near them, until, gathering force, what was at the beginning only a minor movement of small particles becomes a dreaded avalanche, capable not only of sweeping away men in its path but even of obliterating houses and sometimes of changing the whole face of a mountain area. Hence the expression suggested by Ramon y Cajal of the Law of Avalanche for this wide diffusion of sensation, which spreads from a few thousand to millions or billions of cells, and from a rather bearable pain becomes intolerable torture, as a consequence of the brain's complete occupation with it.

Now it is possible for most people, indeed for all who have not some organic morbid condition, to control this spread of pain beyond its original connections, provided only they will to do so, refuse to be ruled by their dreads and proceed to divert attention from the painful condition to other subjects. Here is why the man who bravely faces pain actually lessens the amount that he has to bear. There is no pain in the part affected. That we know, because any interruption of the nerve tract leading from the affected part to the brain eliminates the pain. In the same way, the obtunding of the nerve cells in the cortex by anaesthetics or of the conducting nerve apparatus on the way to the brain by local anaesthesia, will have a like effect. Anything then that will interfere with the further conduction of the pain sensation and the cortical cells directly affected will lessen the sense of pain, and this is what happens when a man settles himself firmly to the thought that he will not allow himself to be affected beyond what is the actual reaction of the nerve tissues to the part.

As a matter of fact, the anticipation of pain due to the dread of it predisposes the part to be much more sensitive than it was before. We can all of us readily make experiments which show this very clearly. Ordinarily we have a stream of sensations flowing up from the surface of the body to the brain, consequent upon the fact that the skin surface is touched by garments over most of the body, and that our nerves of touch respond to their usually rather rough surface. We have learned to pay no attention to these because we have grown accustomed to them, though any one who thinks that they are negligible should witness the writhings of a poor Indian under the stress of being civilized when he is required to wear a starched shirt for the first time. Ordinarily Indians have learned to suppress their feelings, but the shirt with its myriad points of contact, all of them starchily scraping, usually proves too much for his equanimity, and he wiggles and twists to such an extent as shows very clearly that he is extremely uncomfortable. Most people have something of the same feeling the first day that they change into woolen underclothes after they have been wearing cotton for months, and the sensation is by no means easy to bear with equanimity.

Ordinarily from custom and habit in the suppression of feelings we notice none of these contact sensations with their almost inevitable itchy and ticklish feelings, though they are constantly there, but we can reveal them to ourselves by thinking definitely about any part of the body. Such concentration of attention at once brings that part of the body above the threshold of consciousness, and we have distinct feelings there that we did not notice before. If for instance we think about the big toe on the left foot, immediately our attention is turned to it and we note sensations in it that were quite unnoticed before. We can feel the stocking touching any part of it that we think of. Not only that, but if we concentrate attention on a part most uncomfortable sensations develop. If anything calls our attention even to the middle of our backs, we find at once that there is a distinct sensation there, and this may become so insistent as to demand relief.

It is well understood now what happens in these cases. As we have said, the attention given to a part leads to a widening of the minute blood vessels located there so that the nerve endings to the part are supplied with more blood and therefore become more sensitive. We know from experience in cold windy weather that when the cheek is hyperaemic the drawing of a leaf or even of a piece of paper across it may produce a very acute painful sensation. Hyperaemia always makes parts of the body much more sensitive than before. Attention has just this effect over all the surface of the body, as we can demonstrate to ourselves. We can actually, though only gradually, make our feet warm by thinking about them, because the active attention to them sends more blood to them. The dread of pain then, by concentrating attention on the part beforehand, actually increases the pain that has to be suffered and makes the subject ever so much more sensitive. Sensitiveness is of course dependent on other factors, as for instance lack of outdoor air and of oxygenization, which actually seems to hypersensitize people so that even very slight pain becomes extremely difficult to bear, but the question of attention, which is after all almost entirely a voluntary matter, has more to do with making pain harder to bear than anything else.

In the preanaesthetic days, men have been known to sit and watch calmly an amputation of one of their limbs without wincing and apparently without undergoing very much pain. Many are the incidents in history of a favorite general who showed his men how to bear pain by calmly smoking a cigar while a surgeon amputated an arm or a leg or performed some other rather important surgery. Pain is after all like the sense of danger and may be suppressed practically to as great a degree. Once during the present war, when long columns of soldiers going to the front had to pass by the open market place of a town that was being shelled by the Germans, there was danger of the troops losing something of their morale at this point and of confusion ensuing. It would have been disturbing both to discipline and the ordered movement of the troops to divert them by narrower streets, and the shells, though dangerous, were not falling frequently and not working serious havoc. Every one knew, however, that the German gunners had the range, and a shell might land square in the market place at any time; thus there was a feeling of uneasiness and a tendency to nervous lack of self-control, with the inevitable confusion of movement afterwards. One of the French generals ordered an armchair to be brought out of one of the houses near by, took a position in the center of the square, with a little wand in his hand, and calmly joked with the soldiers as they went by about the temperature of the day mentioning occasionally something about a shell that happened to strike not far away. According to the story he was an immense man weighing nearly three hundred pounds, and so provided a very good-sized target for shells, but he was never touched and, almost needless to say, the line of soldiers never wavered while their general sat there joking at the danger.

It is sometimes thought that men in the older, less refined times could stand pain and suffering generally much better than our generation which is supposed to have degenerated in that respect. We have found, however, during the war that the soldiers who could stand supreme suffering the best were very often those who came from better-to-do families, who had been subjected to the most highly refining influences of civilization, but also to that discipline of the repression of the emotions which is recognized as an important phase of civilization. Strange as it may seem, the city boys stood the hardships and the trials of trench life better than the country boys and not only withstood the physical trials but were calmer under fire and ever so much less complaining under injury. After all it is what might be expected, once serious thought is given to the subject, and yet somehow it comes as a surprise, as if the country boy ought to be less sensitive, as indeed he probably is; but he lacks that training in self-control which enables the city boy to stand suffering.

All our feeling that human nature has degenerated in physical constitution has been completely contradicted by the reaction of our young soldiers to camp and trench life. They have gone back to the lack of comforts and conveniences of the pioneer days and have had to submit to the outdoor life and the hardships that their pioneer grandfathers went through and have not failed under them. The boys have come out of it all demonstrating not only that their courage was capable of supporting them, but with their physical being bettered by the conditions and their power to stand suffering revealed in a way that would scarcely have been believed possible beforehand.

CHAPTER IX

THE WILL AND AIR AND EXERCISE

"And wishes fall out as they are willed."
Pericles

Very probably the most important function of the will in its relation to health is that which concerns its power to control the habits of mankind as regards air and exercise. It is surprising to what an extent people neglect both of these essentials of healthy living in the midst of our modern sophisticated life, unless the will power is consciously used for the purpose of forming and then maintaining habits with regard to these requisites for health. It is a very fortunate thing that instinct urges the child, particularly the infant, to almost constant movement during its waking hours. Children that are healthy and that are growing rapidly, boys somewhat more than girls, are so constantly in movement that one would almost think that they must be on springs. Whenever they discover that they can make a new movement, they proceed to make it over and over again until they can do it with facility. There is no lolling around for them; as soon as they wake, they want to be up and doing, no matter what the habits of the household may be. They are constantly on the move. We know that this is absolutely essential for growth as well as for the proper training of their muscles, but it is a very fortunate thing that children do it for themselves, for if their mothers were compelled to train them, the task would be indeed difficult. All mother has to do is to control them to some extent and keep them from venturing too far, lest they should hurt themselves.

When the control of instinct over life is gradually replaced by reason, this tendency to exercise gradually diminishes until it is often surprising to find how little people are taking. As it is mainly the need for exercise that forces people out into the air, indoor life comes to be the main portion of existence. This is all contrary to nature, and so it is not surprising that disease, in its original etymological sense of discomfort, develops rather readily. The lack of exercise in the air permits a great many people to drift into all sorts of morbid conditions in which they are quite miserable. This is, of course, particularly true as regards nervous ailments of various kinds; only under the term nervous ailments should be included not alone direct affections of the nervous system or functional disturbances of nerves, but also a number of other conditions. Nervous indigestion, insomnia, neurotic constipation and many of the symptomatic affections associated with these conditions, tired feelings that interfere with activities, headache, various feelings of discomfort in the muscles and around the joints, inability to control the emotions and other such common complaints if that is the proper word for them all these are fostered by a sedentary life indoors. They frequently make not only the patient himself or oftener herself miserable, but also all those who come in contact with her.

Above all, it must not be forgotten that lack of exercise in the open air has a very definite tendency to make people extremely sensitive to discomforts of all kinds, mental as well as physical. Many a man or woman whose life seems full of worries, sometimes without any adequate cause at all, who goes from one dread to another, who wakes in the morning with a sense of depression, find that most of these feelings and sometimes all of them, disappear promptly when they begin to exercise more in the open.

Nothing dispels the gloom and depressions consequent upon an accumulation of cares and worries of various kinds like a few weeks in the woods, where every moment is passed in the fresh outdoor air, which actually seems to blow the cobwebs of ill feelings away and leaves the individual with a freedom of mind and a comfort of body that he almost expected never to enjoy again.

Undoubtedly the most important factor for the preservation of health is an abundance of fresh air. At certain seasons of the year this is not only easy and agreeable, but to do anything else imposes hardship. In our climate, however, there are about six months of the year in which it requires some exercise of will power to secure as much open air life as is required for health. There are weeks when it is too hot, there are many weeks when it is too cold. The cold air particularly is important, because it produces a stimulating vital reaction than which nothing is more precious for health. We have no tonic among all the drugs of the pharmacopeia that is equal to the effect of a brisk walk in the bracing air of a dry cold day. After a long morning and perhaps a whole day in the house, even half an hour outdoors will enable us to throw off the sluggishness consequent upon confinement to the indoor air and the lack of appetite and the general feeling of physical lassitude which has followed living in an absolutely equable temperature for twenty-four hours. Sometimes it requires no little effort of the will to secure this, and to continue it day after day without missing it or letting it be crowded out by claims that are partly real and partly excuses, because we do not care to make the special effort required.

What humanity needs is regular exercise in the open air every day. As it is, between the trolley car and the automobile, very few people get what they need. Any one who has to go a mile takes a car or some other conveyance and between waiting for the car and certain inevitable delays it will probably take ten minutes or more to go the mile. In five minutes more one could walk that distance and secure precious exercise besides such diversion of mind as inevitably comes from walking on busy city streets and which makes an excellent recreation in the midst of one's work. For it is quite impossible in our day to walk along city streets absorbed in abstract mental occupations. One of the objections to walking is that after a while it can be accomplished as a matter of routine without necessarily taking one's mind away from subjects in which it has been absorbed. It is quite impossible for this to happen, however, on modern city streets. "The outside of a horse", it used to be said, "is good for the inside of a man." The main reason for this was because it is impossible for a man to ride horseback, unless his mount is a veritable old Dobbin, without paying strict attention to the animal. The same thing is true as regards city pedestrianism, especially since the coming of the auto has made it necessary to watch our steps and look where we go.

A great many people would be ever so much better in health if they walked to business or to school every morning instead of riding, for the young need it even more than the older people. Especially is this true for all those who follow sedentary occupations. Clerks in lawyers' offices, typewriters and stenographers, secretaries all those who have to sit down much during the day need the brisk walking and need it not merely of a Sunday or a Saturday afternoon, but every day in the year. Many of them, if they walked two and three miles to the office, would probably require only fifteen minutes, at most half an hour, more than if they took a train or trolley, but they would have secured a good hour of exercise in the open air.

On the other hand the unfortunate crowding of trolley and elevated and subway trains in the busy hours when people go to and from their work makes an extremely uncomfortable and often rather depressing commencement and completion of the day's work. I know of nothing that makes a worse beginning for the day than to have to stand for half an hour or longer in a swaying, bumping car, hanging to a strap, crushed and crowded by people getting in and out. The effect of coming home under such circumstances after a reasonably long day's work is even more serious, and any little sacrifice that will enable people to avoid it will do them a great deal of good. Fifteen or twenty minutes of extra time morning and evening would often suffice for this and would at the same time add a bracing walk in the open air to the day's routine.

When first begun, such a practice would make one tired and sore, but that condition would pass in the course of a few days and be replaced by a healthy feeling of satisfaction that would be well worth all the effort required. We should need ever so much less medicine for appetite and for constipation if this were true. A great many people who stand during the day would probably deem it quite out of the question for them to walk three miles or more to and from their business, for their feet get so tired that they feel that they could not endure it. What they need more than anything else, however, is exercise that will bring about a stimulation of the circulation in their feet. Standing is very depressing to the circulation. It leads to compression of the veins and hence interference with the return circulation, with lowered nutrition which often predisposes to flat foot or yielding arch and tends to create corns and callouses: walking in reasonably well fitting shoes on the contrary tends to make the feet ever so much less sensitive Our soldiers have had that experience and have learned some very precious lessons with regard to the care of their feet, the principal one being that the best possible remedy for foot troubles is to exercise the feet vigorously in walking and running, provided the shoes permit proper foot use.

I have often known clerks and floorwalkers who have to stand all day or move but a few steps at intervals, who were so tired at night that they felt the one thing they could do was to sit down for a while after dinner and then go to bed, but who came to feel ever so much better after a brisk walk home. It was rather hard to persuade them that, exhausted as they felt, they would actually get rested and not more tired from vigorous walking, but once they tried it, they knew the exercise was what they needed. The air in stores is often dry and uncomfortable for those who are in them all day. It is usually and quite properly regulated for the customers who come in from the streets expecting to get warm without delay. In dry, cold weather particularly, an evening walk home sets the blood in circulation until it gets thoroughly oxidized and the whole body feels better. Such a brisk walk will often prevent the development of flat foot, especially if care is taken to spring properly from the ball of the foot, in the good, old-fashioned heel and toe method of walking. Once flat foot has developed, walking probably is more difficult, but even then, with properly fitting shoes, the patients will be the better for a good walk after their work is over. It requires some will power to acquire the habit, but once formed, the benefit and pleasure derived make it easy to keep up the practice.

Those who walk thus regularly will often find that their evening tiredness is not so marked, and they will feel much more like going out for some diversion than they otherwise would. Probably nothing is more dispiriting in the course of time than to come home merely to eat dinner, sit down after dinner and grow sleepy on one's chair until one feels quite miserable, and then go to bed. There should be always, unless in very inclement weather, an outing before bedtime, and this should be looked forward to. It will often forestall the feeling that the day is over after dinner and so keep the individual from settling down into the dozy discomfort of an after-dinner nap as the closing scene of the day. Good habits in this matter require an effort of the will to form; bad habits almost seem to form of themselves and then require a special effort to break.

It is surprising how many of the dreads and anxiety neuroses and psycho-neurotic solicitudes and neurasthenic disquietudes and other more or less morbid mental states disappear under the influence of a brisk walk for three or four miles or more every day. I have tried this prescription on all sorts of people, including particularly myself, and I know for certain that when troubles are accumulating the thing to do is to get outdoors more, especially for walking; then the incubus begins to lift. Clergymen, university professors, members of religious orders, school teachers, as well as bankers, clerks and business people of various kinds, have been subjected to the influence of this prescription with decided benefit. Some of them assert that they never felt so well as since they have formed the habit of walking every day. It must, however, be every day, and it must not merely be a mile or so but it must be at least three miles. That means for a good many people about an hour spent in actual walking, but it is well worth the time and effort. Above all, it repays not only in health and in better feelings but in the increased amount of work that can be done on the day itself. A whole day passed indoors will often contain many wasted hours, while if a walk of a couple of miles is planned for the morning and one for a couple of miles more in the afternoon, very satisfactory study or other work can be done in the intervals. Almost needless to say, a brisk walk in the cooler weather will create an appetite where it did not exist before. Women often need counsel in this matter more than men, and regular walking for them is indeed a counsel of health. Very few women in these modern times walk much, and to walk more than a mile seems to them a hardship. This is responsible for more of the supersensitiveness and nervous complaints of all kinds to which women are liable than anything else that I know of. It is also one important factor in the production of the constipation to which women are so much more liable than men. We see many advertisements with regard to the jolts to which the body is subjected every time the heel is put down and of the means that should be taken to prevent them, but it must not be forgotten that men and women were meant by nature to walk erect and that this recurring jolt has a very definite effect in stimulating peristalsis and favoring the movement of the contents of the intestines. Besides, if the walking is brisk, the breathing is deeper and there is some massage of the liver, as also of the other abdominal viscera, while other organs are affected favorably. Walking for women regular, everyday walking would be indeed a precious habit, but now that women have occupations more and more outside of the house, this is one of the things they must make up their minds to do, if they are to maintain health, remembering that making up the mind is really making up the will.

Over and over again I have seen a great many of the troubles of the menopause or change of life in women disappear or become ever so much less bothersome as the result of the formation of regular habits of walking out of doors every day. Unfortunately, there is a definite tendency about this time for women to withdraw more and more from public appearances and to live to a considerable extent in retirement at home. Nothing could be much worse for them. They need, above all, to get out and to have a number of interests, and if these interests can only be so arranged as to demand rather prolonged walks, so much the better. This is more particularly true for the unmarried woman who is going through this critical time, and the question of walking regularly every day for three or four miles must be proposed to her. It will require a considerable effort of the will. More than two miles at the beginning will probably be too tiring, but the amount can be gradually increased until at least four miles on the average is covered every day. Above all, for the feelings of discomfort in the cardiac region so often noticed at this time, regular walking is the best remedy in most cases, always of course presupposing that there is no organic heart condition, for in that case only a physician can give the proper direction for each case. By the exercise of the lungs that it requires, it will probably save most people from colds and coughs which they have had to endure every winter. Lastly be it said that practically all men and women, though more particularly the men who have lived well beyond the Psalmist's limit of threescore and ten, have been regular daily walkers, or else they have taken exercise in some form in the open air which is the equivalent of walking. One of the most distinguished of English physicians, Sir Hermann Weber, who died just after the end of the war in London, was in his ninety-fifth year. He had practised medicine regularly until the age of eighty and continued in excellent health and vigor until just before his death. During the last year of life, he contributed an interesting article to the British Medical Journal on the "Influence of Muscular Exercise on Longevity." He attributed his vigor at the age of ninety-five as well as the prolongation of his life to his practice of spending every day two or three hours in the open air. He walked, as a rule, forty to fifty miles a week. Even in the most inclement weather he rarely did less than thirty miles a week. Many another octogenarian and nonagenarian has attributed his good health and long life to the habit of regular daily exercise in the open.

Instead of using up energy, the will so used brings out latent stores of energy that would not otherwise be employed and thus adds to the available amount of vitality for the individual. Doctor Thomas Addis Emmet, only just dead, over ninety, in his younger years as a busy medical practitioner never kept a horse. It would not be difficult to cite many other examples among men who lived to advanced old age and who considered that they owed their good health and long life to daily habits of outdoor exercise.

CHAPTER X

THE WILL TO EAT

"If your will want not, time and place will be fruitfully added."
King Lear

Eating is usually supposed to be entirely a matter of appetite which instinct directs to the best possible advantage of the individual. This is quite true for those who are living the outdoor life that is normal or at least most healthy for men, and when they are getting an abundance of exercise, and may I add also have not too great a variety of food materials in tempting form presented to them. Under the artificial not to say unnatural conditions which men have to a great extent created for themselves in city life, confined at indoor sedentary occupations, some of them and they are much more numerous than is usually imagined eat too little, while a great many, owing to stimulation of appetite in various ways, eat too much.

Eating therefore for health's sake has to be done through the will and as a rule by the formation of deliberate habits. It is easy to form habits either of defect or excess in the matter of eating and indeed a great deal of the ill health to which mankind is liable is due to errors in either of these directions. Having disturbed nature's instincts for food in modifying the mode of life to suit modern conveniences, we have now to learn from experience and scientific observations what we should eat and then make up our minds to eat such quantity and variety as is necessary to maintain health and strength in the particular circumstances in which we are placed.

While the greatest emphasis has been placed on the dangers to health in overeating, the number of people who, for one reason or another, eat too little is, as has been said, quite surprising. A very large proportion of those under normal weight are so merely because they have wrong habits of eating. Indeed, it may be laid down as a practical rule of health that wherever there is no organic disease the condition of being underweight is a symptom of undereating. A great many thin people insist that the reason why they are underweight is that it is a family trait and that father and mother, or at least one of them, and some of their grandparents exhibited this peculiarity; and thus it is not surprising that they should have it. A careful analysis of the family eating in such cases has shown me in a large number of instances, indeed almost without exception, that what my patients had inherited was not a constitutional tendency to thinness, but a family habit of undereating. This accrued to them not from nature but from nurture, and was acquired in their bringing up. Most of them were eating one quite abundant meal a day and perhaps a pretty good second meal, but practically all of them were skimping at least one meal very much. In some way or other, a family habit of eating very little at this meal had become established and was now an almost inviolable custom.

A great many thin individuals, that is persons who are somewhat more than ten per cent. under the average normal weight for their height, either do not eat breakfast at all or eat a very small one. It is not unusual for the physician analyzing their day's dietary to be told that the meal consists of a cup of coffee and a piece of bread. Sometimes there is a roll, but more often only part of a roll, though occasionally in recent years there may be some fruit and some cereal; the fruit will usually be a half of one of the citrus fruits which contains practically no nutrition and is only a pleasant appetizer, while more often than not the cereal will be one of the dry, ready-to-eat varieties which, apart from the milk or cream that may be served with them, contain in the usual small helpings very little nutriment. Such breakfasts are particularly the rule among women who are under weight. Sometimes lunch is comparatively light so that there are two daily apologies for meals. To make up for these, the third meal may be very hearty. City folk often eat at dinner more than is good for them. This may produce a sense of uncomfortable distention and overfulness followed by sleepiness which may be set down as due to indigestion, though it is just a question of overeating for the nonce.

It would be much more conducive to health to distribute the eating over the three meals of the day, but it requires a special effort of the will to break the unfortunate habits that have been formed. Particularly it seems hard for many people to eat a substantial breakfast and a determined effort is required to secure this. It would seem almost as though their wills had not yet waked up and that it was harder for them to do things at this time of day. It is especially important for working women, that is, those who have such regular occupations as school-teacher, secretary, clerk and the like, to eat a hearty breakfast. They can get a warm properly chosen meal at home at this hour, while very often in the middle of the day they have to eat a lunch that is not nearly so suitable. As a consequence of neglecting breakfast then, it is twenty-four hours between their warm, hearty meals. Even when they eat a rather good lunch, some eighteen hours elapse since the last hearty meal was taken, and one half the day's work has to be done on the gradually decreasing energy secured from the evening meal of the day before. With this unfortunate habit of eating, most of that was used up during the night in repairing the tissue losses of the day before, so that the morning's work has to be done largely "on the will" rather than on the normal store of bodily energy.

It is surprising how many patients who are admitted to tuberculosis sanatoria have been underweight for years as a consequence of unfortunate habits of eating. Not infrequently it is found that they have a number of prejudices with regard to the simple and most nutritious foods that mankind is accustomed to. Not a few of the younger ones who develop tuberculosis have been laboring under the impression that they could not digest milk or eggs or in some way they had acquired a distaste for them and so had eliminated them from their diet; some of them had also stopped eating butter or used it very sparingly. At the sanatoria, as a rule, very little attention is paid to the supposed difficulty of digestion of milk and eggs and perhaps butter. The patients are at once put on the regular diet containing these articles and the nurse sees that they take them even between meals, and unless there is actual vomiting or some very definite objective not merely subjective sign of indigestion, the patients are required to continue the diet.

It is almost an invariable rule for the patients of such institutions to come to the physician in charge after a couple of weeks and ask how it was that they could have thought that these simple articles of food disagreed with them. They have begun to like them now and are surprised at their former refusal to take them, which they begin to suspect, as the physician very well knows, to have been the principal reason for the development of their tuberculosis.

There are people who are up to weight or slightly above it who develop tuberculosis, but they do not represent one in five of the patients who suffer from the affection. In probably three fourths of all the cases of tuberculosis the predisposing factor which allowed the tubercle bacillus to grow in the tissues was the loss of weight or the being underweight. There is a good biological reason for this, for there are certain elements in the make-up of the tubercle bacillus which favor its growth at a time when fat is being lost from the tissues rather than deposited, for at that time more fat for the growth of the tubercle bacillus is available in the lungs than at other times. Often among the poor the loss in weight is due to lack of food because of poverty, or failure to eat because of alcoholism, but not infrequently among all classes it is just a question of certain bad habits of eating that might readily have been corrected by the will. It is surprising how many people who complain of various nervous symptoms meaning by that term symptoms for which no definite physical basis can be found, or for which only that extremely indefinite basis of a vague reflex, real or supposed, from the abdominal organs are underweight and will be found to be eating much less than the average of humanity. These nervous symptoms include above all discomforts of various kinds in the abdominal region; sense of gone-ness; at times a feeling of fullness because of the presence of gas; grumblings, acid eructations, bitter taste in the mouth, and above all, constipation. As is said in the chapter on "The Will and the Intestinal Functions," the most potent and frequent cause of constipation is insufficient eating, either in quantity or in variety. It is especially in the digestive tract of those who do not eat as much as they should that gas accumulates. This gas is usually thought to be due to fermentation, but as fermentation is a very slow gas producer and nervous patients not infrequently belch up large quantities, it is evident that another source for it must be sought. Any one who has seen a number of hysterical patients with gaseous distention of the abdomen and attacks of belching in which immense quantities of gas are eructated, will be forced to the conclusion that in such nervous crises gas leaks out of the blood vessels of the walls of the digestive tract and that this is the principal source of the gas noted. What is true in the severe nervous attacks is also true in nervous symptoms of other kinds, and neurotic indigestion so called is always accompanied by the presence of gas.

Apparently the old maxim of the physicist of past centuries has an application here. "Nature abhors a vacuum" and as the stomach and intestines are not as full as they ought to be, nor given as much work to do as they should have, nature proceeds to occupy them with gas which finds its way in from the very vascular gastrointestinal walls. This is of course an explanation that would not have been popular a few years ago when the chemistry of digestion seemed so extremely important, but in recent years, medical science has brought us back rather to the physics of digestion, and I think that most physicians who have seen many functional nervous patients would now agree with these suggestions as to the origin of gaseous disturbance in the gastrointestinal tract in a great many of these cases.

Besides the physical symptoms, there are a whole series of psychic or psycho-neurotic symptoms, the basis of which undoubtedly lies in the condition of underweight as a consequence of undereating. Over and over again I have seen the feeling of inability to do things which had come over men, and particularly women, disappear by adding to and regulating the diet until an increase in weight came. Extreme tiredness is a frequent symptom in those under weight, and this often leads to their having no recreation after their work because they have not enough energy for it; as every human being needs diversion, a vicious circle of influence which adds to their nervous tired condition is formed. I have seen in so many cases the eating of a good breakfast and a good lunch supply working people with the energy hitherto lacking that enabled them to go out of an evening to the theater or to entertainments of one kind or another, that it has become a routine practice to treat these people by adding to their dietary unless there are direct contra-indications.

Dreads are much more common among people who are underweight than among those who eat enough to keep themselves in proper physical condition. I have had a series of cases, unfortunately only a small one in number, in which the craving for alcoholic liquor disappeared before an increase in diet and a gain in weight. I shall never forget the first case in which this happened. The patient was a man of nearly sixty years of age who held a rather important political office in a small neighboring town. He was on the point of losing it because periodical sprees were becoming more frequent and it was impossible for him to maintain his position. He was over six feet in height and he weighed less than a hundred and fifty pounds. I had tried to get him to gain in weight by advice and suggestion without avail. Finally, I had to make a last effort to use whatever influence I had to save his political position for him, and then I succeeded in making him understand that he would have to do as I told him in the matter of eating, or else I would have nothing more to do with him.

It was not without some misgivings that I thus undertook to make a man of nearly sixty change his lifelong habits of eating. That is something which I consider no physician has a right to do unless there is some very imperative reason for it. Here was, however, a desperate case. It was in the late afternoon particularly that this patient craved drink so much that he could not deny himself. As he ate but very little breakfast and had a hasty scanty lunch, he was at the very bottom of his physical resources at that time, and at the end of a rather demanding day's work. We had to break up his other habits in the hope of getting at the craving. He had taken coffee and a roll for breakfast. I dictated a cereal, two eggs and several rashers of bacon and several rolls. I insisted on fifteen minutes in the open before lunch and then a hearty lunch with some substantial dessert at the end of it. This man proceeded to gain at the rate of a little more than three pounds a week. By the end of two months, he weighed about one hundred and eighty pounds and had not touched a drop of liquor in that time and felt that he had no craving for it. That is some ten years ago, and there has been no trouble with his alcoholic cravings since. He has maintained his weight; he says that he never felt so well and that above all he now has no more of that intense tiredness that used to come to him at the end of the day. Every now and then he says to me in musing mood, "And to think that I had never learned to eat enough!"

For these very tired feelings so often complained of by nervous patients, once it has been decided that there is no organic trouble for of course kidney or heart or blood pressure affections may readily cause them there are just two things to be considered: These are flat-foot or yielding arch, and undereating. When there is a combination of these two, then tiredness may well seem excessive and yet be readily amenable to treatment. Persons with occupations which require standing are especially liable to suffer in this way.

Undereating in the evening is especially important for many nervous people and is often the source of wakefulness. It is the cause of insomnia, not so much at the beginning of the night, as a rule, as in the early morning. Many a person who wakes at four or five and cannot go to sleep again is hungry. There is a sense of gone-ness in the stomach region in these cases, which the patients are prone to attribute to their nerves in general, or some of them who have had unfortunate suggestions from their physicians may talk of their abdominal brain; but it is surprising how often their feelings are due simply to emptiness. Any thin person particularly who has his last meal before seven and does not go to bed until after eleven should always take something to eat before retiring. A glass of milk or a cup of cocoa and some crackers or a piece of simple cake may be sufficient, but it is important to eat enough. Animals and men naturally get sleepy after eating and do not sleep well if their stomachs are empty. Children are the typical examples. We are all only children of a larger growth in this regard.

When the last meal is taken before seven and people do not go to bed until nearly twelve, as is frequently the case in large cities, the custom of having something to eat just before bed is excellent for sleep. I have known the establishment of this habit to afford marked relief in cases of insomnia that had extended over years. The people in my experience who sleep the worst are those who, having taken a little cambric tea and some toast and preserves with perhaps a piece of cake for supper, think that this virtuous self-control in eating ought to assure them good rest. It has just the opposite effect. Disturbed sleep, full of dreams and waking moments, is oftener due to insufficient eating than to overeating. The people whom I know who sleep the best and from whom there are no complaints of insomnia, are those who, having eaten so heartily at dinner that they get to the theater a little late, attend the Follies or some late show for a while and then go round to one of the Broadway restaurants and chase a Welsh rarebit or some lobster a la Newburg, with a biscuit Tortoni or a Pêche Melba down to their stomachs and then go home to sleep the sleep of the just.

Just as there are bad habits of eating too little that are dangerous and must be corrected by the will so there are bad habits of eating too much that can only be corrected in the same way. While it is dangerous to be under weight in the early years of life, it is at least as dangerous to be overweight in middle life. With the variety and abundance of food now supplied at a great many tables, it is comparatively easy for people in our time to eat too much. The result is that among the better-to-do classes a great many people suffer from obesity, sometimes to such an extent that life is made a burden to them. There is only one way to correct this and that is to eat less and of course to exercise more. Reduction in diet means the breaking of a long established habit and that of course is often hard. The whole family may have to set a good example of abstinence from too great a variety of food and especially from the richer foods, in order that a parent may be helped to prevent further development of obesity and to lose gently and gradually some of the overweight that is being put on, and which now, by conserving heat and slowing up metabolism generally within the body, makes it so easy for even reduced quantities of food to maintain the former habit of adding weight.

In this matter of obesity, however, just exactly as in the case of tuberculosis for those who are underweight, prevention is much better than cure. The people who know that they inherit such tendencies should be particularly careful not to form habits of eating that will add considerably to their weight. After all, it is not nearly so difficult a matter as is often imagined. There is no need, unless in very exceptional cases, of denying one's self anything that is liked in the ordinary foods, only less of each article must be eaten. Even desserts need not be entirely eliminated, for ices may be taken instead of ice cream; sour fruits and especially those of the citrus variety oranges and grapefruit and the gelatine desserts may be eaten almost with impunity. The phrase "eat and grow thin" has deservedly become popular in recent years because as a matter of fact it is perfectly possible to eat heartily and above all to satisfaction without putting on weight. It is, of course, harder to lose weight, but even that may be accomplished gradually under proper direction if there is the persistent will to do it.

In recent years another disease has come to attract attention which represents the result of an overindulgence in food materials that can be limited without much difficulty. This is diabetes which used to be comparatively rare but has now become rather frequent. An authority on the disease declared not long since that there are over half a million people in this country now who either have or will have diabetes as the result of the breaking down of their sugar metabolism. It is not surprising that the disease should be on the increase, for the consumption of sugar has multiplied to a very serious degree during the last few generations. A couple of centuries ago, those who wanted sugar went not to the grocery store, but to the apothecary shop. It was kept as a flavoring material for children's food, as a welcome addition to the dietary of invalids and the old, and quite literally as a drug, for it was considered to have, as it actually has, to a slight extent at least, some diuretic qualities that made it valuable. A little more than a century ago, a thousand tons of sugar sufficed for the whole world's needs, while the year before the war, the world consumed some twenty-two million of tons of sugar. It is said that every man, woman, and child in the United States consumed on the average every day a quarter of a pound of sugar.

Our candy stores have multiplied, and while two generations ago the little candy stores sold candies practically entirely for children, eking out their trade with stationery and newspapers and school supplies, now candy stores dealing exclusively in confectionery are very common. There are several hundred stores in the United States that pay more than $, a year rent, though they sell nothing but candy and ice-cream sodas. Corresponding with the increase in the sale of candy has come also the consumption of very sweet materials of various kinds. French pastries, Vienna tarts, Oriental sweetmeats, Turkish fig paste, Arabian date conserves, and West Indian guava jelly, are all familiar products on our tables. Chocolate has become one of the important articles of world commerce, though almost unknown beyond a very narrow circle a little more than a century ago. Tea and coffee have been introduced from the near and the far East and by a Western abuse consumed with such an amount of sweetening as make them the medium of an immense consumption of sugar.

There is no doubt that unless good habits of self-denial in this regard are formed, diabetes, which is an extremely serious disease, especially for those under middle life, will continue to increase in frequency. The candy and sugar habit is rather easy to form; every one realizes that it is a habit, but it is sometimes almost as hard to break as the tobacco habit. We were meant to get our sugar by the personal manufacture of it from starch substances. If a crust of bread is chewed vigorously until it swallows itself, that is, dissolves in the secretions and gradually disappears, it will be noted that there is a distinctly sweetish taste in the mouth. This is the starch of the bread being changed into sugar. We were expected by nature to make our own sugar in this way, but this has proved too slow and laborious a way for human nature to get all the sugar it cared for, so most people prefer to secure it ready made. Sugar is almost as artificial a product as alcohol and is actually capable of doing almost as much harm as its not distantly related chemical neighbor. It is rather important that good habits in the matter should be formed and we have been letting ourselves drift into very unfortunate habits in recent years.

CHAPTER XI

THE PLACE OF THE WILL IN TUBERCULOSIS

"And like a neutral to his will and matter
Did nothing."
Hamlet

Probably the very best illustration in the whole range of medicine of the place of the will in the cure of disease is afforded by tuberculosis. This used to be the most fatal of all human affections until displaced from its "bad eminence" within the last few years by pneumonia, which now carries off more victims. As it is, however, about one in nine or perhaps a few more of all those who die are victims of tuberculosis. This high mortality would seem to indicate that the disease must be very little amenable to the influence of the will, since surely under ordinary circumstances a good many people might be expected to have the desire and the will to resist the affection if that were possible. In spite of the large death rate this is exactly what is true.

Tuberculous infections are extremely common, much commoner even than their high mortality reveals. After long and critical discussion with a number of persistent denials, it is now generally conceded by authorities in the disease that the old maxim "after all, all of us are a little tuberculous" is substantially correct. Very few human beings entirely escape infection from the tubercle bacillus at some time in life. The great majority of us never become aware of the presence of the disease and succeed in conquering it, though the traces of it may be found subsequently in our bodies. Careful autopsies reveal, however, that very few even of those who did not die directly from tuberculosis fail to show tuberculous lesions, usually healed and well shut off from the healthy tissues, in their bodies. One in eight of those who become infected have not the resistive vitality to throw off the disease or the courage to face it and take such precautions as will prevent its advance. All those, however, who give themselves any reasonable chance for the development of resistance survive the disease though they remain always liable to attack from it subsequently if they should run down in health and strength.

Heredity, which used to be supposed to play so important a rôle in the affection, is now known to have almost nothing to do with the spread of the disease. Family tendencies are probably represented by nothing more than a proneness to underweight which makes one more liable to infection, and this is due as a rule to family habits in the matter of undernourishment from ill-advised consumption of food. Probably a certain lack of courage to face the disease boldly and do what is necessary to develop bodily resistance against it may also be an hereditary family trait, but environment means ever so much more than heredity.

There is a well known expression current among those who have had most experience in the treatment of patients suffering from tuberculosis that "tuberculosis takes only the quitters", that is to say that only those succumb to consumption who have not the strength of will to face the issue bravely and without discouragement to push through with the measures necessary for the treatment of their disease. In a word it is only those who lack the firmness of purpose to persist in the mode of life outlined for them who eventually die from their affection of the lungs. No specific remedy has been found that gives any promise of being helpful, much less of affording assured recovery, though a great many have been tried and not a few are still in hopeful use. Recent experience has only served to emphasize the fact that the one thing absolutely indispensable for any successful treatment of tuberculosis of the lungs is that the patient should regain weight and strength and with them resistive vitality so as to be able to overcome the disease and get better.

To secure this favorable result two conditions of living are necessary but they must be above all persisted in for a considerable period. First there must be an abundance of fresh air with rest during the advancing stage or whenever there are acute symptoms present, and secondly an abundance of good food which will provide a store of nutritive energy and make the resistive vitality as high as possible. Curiously enough this "fresh air and good food" treatment for the disease was recognized as the sheet anchor of the therapeutics of consumption as long ago as Galen's time, the end of the second century, when that distinguished Greek physician was practising at Rome. Nearly eighteen hundred years ago Galen suggested that he had tried many remedies for what he called phthisis, the Greek equivalent of our word consumption or wasting away, and had often thought that he had noted a remedial value in them, but after further experience he felt that the all-important factors for cure were fresh air and good food. He even went so far as to say that he thought the best food of the consumptive or the phthisical, as he called them, was milk and eggs. A great deal of water has flowed under the bridge of medical advance since his time and at many periods since physicians have been sure that they had valuable remedies for consumption; yet here we are practically back at Galen's conclusion more than fifty generations after his time, and we are even inclined to think of this mode of treatment as comparatively new, as it is in modern history.

The influence on consumption of the will to get well when once aroused was typically exemplified in the career of the well-known London quack of the beginning of the nineteenth century, St. John Long. He set himself up as having a sure cure for consumption. He was a charlatan of the deepest dye whose one idea was to make money, and who knew nothing at all about medicine in any way. He took a large house in Harley Street and fitted it up for the reception of people anxious to consult him. For some seasons every morning and afternoon the public way was blocked up with carriages pressing to his door. Nine out of ten of his patients were ladies and many of them were of the highest rank; fashion and wealth hastened to place themselves and their daughters at the mercy of the pretender's ignorance. His mode of treatment was by inhalation. He assured his patients that the breathing in of this medicated vapor would surely cure their pulmonary disease, and because others were intent on going they went; many of them were greatly benefited for a time and these so-called cures proved a bait for many other patients.

J. Cordy Jeaffreson in his volume "A Book about Doctors", written two generations ago, has told the story of St. John Long's successful application of the principle of community of treatment and its effectiveness upon his patient. Like Mesmer he realized that treating people in groups led them mutually to influence each other and to bring about improvement. St. John Long had in one of the rooms in Harley Street "two enormous inhalers, with flexible tubes running outward in all directions and surrounded by dozens of excited women ladies of advanced years and young girls giddy with the excitement of their first London season puffing from their lips the medicated vapor or waiting until a mouthpiece should be at liberty for their pink lips." In our generation of course we had various phases of similar treatment, including nebulizers and compressed air apparatus and medicated vapor, all working wonders for a while, and then proving to have no physical beneficial effect.

What is surprising is to find the number of cures that were worked. St. John Long had so many applicants for attention that he was literally unable to give heed to all of them. The news of the wonderful remedy flew to every part of the United Kingdom and from every quarter sick persons, wearied of a vain search after an alleviation of their sufferings, flocked to London with hope renewed once more. This enabled St. John Long to select for treatment only such cases as gave ready promise of cure. He made it a great preliminary of his treatment that his patients should eat well as a rule and on one occasion when he was called into the country to see a man suffering in the last stages of consumption he said quite frankly, "Sir, you are so ill that I cannot take you under my charge at present. You want stamina. Take hearty meals of beefsteak and strong beer; and if you are better in ten days I will do my best for you and cure you."

It is easy to understand that if he made it a rule for his consumptive patients that they should eat well or not expect relief from his medicine he would secure a great many good results. Especially would this be true in many cases that came up to him from the country, had the advantage of a change of climate, and of environment and very soon found that they had much more strength than they thought they had. They had been dreading the worst, they were now led to hope for the best; they took the brake off their will, they fed well and it was not long then before they proceeded to get well.

As even a little experience with consumptive patients shows it is often difficult for them to follow directions and keep it up in the matter of fresh air and good food and here is where the question of the will in the treatment is all important. Many a consumptive has in early life formed bad habits with regard to eating, especially in the direction of eating too little and refusing for some reason or other to take what are known to be the especially nutritious foods. Not infrequently indeed it is their neglect of nutrition in this regard that has been the principal predisposing factor toward the development of the disease. This bad habit must be overcome and often proves refractory.

Then it is never easy to give up the pursuit of a chosen vocation and pursue faithfully for a suitable period the humdrum monotonous existence of prolonged rest every day in the open air with eating and sleeping as almost the only serious interests, if indeed they can be called such, permitted in life. It is only those who have the will power to follow directions faithfully, whole-heartedly and persistently who have a reasonable prospect of getting ahead of their disease and eventually securing such a conquest of it as will enable them to return to their ordinary life as it was before the development of tuberculosis.

Unless patients are ready to follow directions as regards outdoor air and good food the cure, or as specialists in tuberculosis prefer to call it the arrest of symptoms in the disease, is almost out of the question. Above all it is extremely important that those who suffer from pulmonary tuberculosis should be ready to follow directions at an early stage of their disease, before any serious symptoms develop, for it is then that most can be done for them. Many a sufferer from tuberculosis makes his or her cure extremely difficult, certainly ever so much more difficult than it would otherwise have been, because the dread of going to see a physician lest they should be told that their affection is really consumption and demands immediate strenuous treatment causes them to put off consultation with some one whose opinion in the matter is reliable.

This is indeed one of the principal reasons why tuberculosis of the lungs still continues to carry off so many victims every year, because people are afraid to learn the truth. They dare not put the question to a definite issue and refuse to believe the possibility that certain disturbing symptoms represent developing tuberculosis. They defer seeing an expert; they take this and that suggestion from friends; they buy cough remedies which they see advertised, sometimes they tinker with so-called "consumption cures." After a while an advance of their symptoms makes it absolutely necessary to see a physician but often by this time their disease has progressed from an incipient case rather easy to be treated and with an excellent prognosis to a more advanced stage at which cure is ever so much more difficult; or by this time it may even prove that their strength has been seriously sapped and they have not enough resistive vitality left to bring about reaction toward the cure.

The all-important thing for all those who have at any time lived near consumptives, whether relatives or others for the disease is almost invariably acquired and not hereditary or who have worked for any prolonged period in more or less intimate contact with those who had a chronic cough or who subsequently developed tuberculosis, is that on the first symptom that is at all suspicious they should make up their minds to have the question as to whether they have tuberculosis or not definitely settled and that they should be ready to do what they are told in the matter. The first symptom is not a persistent cough as so many think, nor continued loss of weight, which is an advanced sign as a rule, but a continued rapidity of pulse for which no non-pulmonary reason can be found.

The old idea that consumptives should not be told what their affection was, lest it should disturb their minds and discourage them so much as to do them harm, has now been abandoned by practically all those of large experience in the care of the tuberculous. The opposite policy of being perfectly candid and making the patients understand their serious condition and the importance of taking all the measures necessary for cure, yet without permitting them to be unnecessarily scared, has been adopted. Their will to get well must be thoroughly aroused. After all, it must be recalled that tuberculosis is an extremely curable disease. It is now definitely known that more than ninety per cent. of humanity have at some time had a tuberculosis process, that is to say a focus of tuberculosis active within their tissues. Only about one in nine of the deaths in civilized countries is from tuberculosis. That means that at least eight other people who have not died from the disease but from something else have had the affection, yet have recovered from it. Instead of the old shadow of heredity with its supposedly almost inevitable fatality, so that young people who saw their brothers and sisters or other relatives around them die from the disease felt that they were doomed, we now know that the hereditary factor plays an extremely minor role if indeed it plays any serious rôle at all in the development of the disease.

No affection is so amenable to the state of mind and the will to be well as tuberculosis. That is exactly the reason why so many remedies have come into vogue and apparently been very successful in its treatment and then after a while have proved to be of no particular service or even perhaps actually harmful so far as their physical effect is concerned. It cannot be too often repeated that anything whatever that a patient takes that will arouse new hope and give new courage and reawaken the will will actually benefit these patients. No wonder then that scarcely a year passes without some new remedy for tuberculosis being proposed. All that is needed to affect favorably patients suffering from the disease is to have some good reason presented which makes them feel that they ought to get better and then at once they eat better and proceed to increase their resistive vitality. The despondency that comes with the lack of the will to be well hurts their appetite particularly and no tuberculosis patient can ever hope to recover health unless he is eating heartily. With better eating there is always a temptation to be more outdoors and the ability to stand cooler air which always means that the lungs are given their opportunity to breathe fresh cool air which constitutes absolutely the best tonic that we have for the affection.

It has been recognized in recent years that the only climates which give reasonable hope of being helpful for the tuberculous are those which present a variation of some thirty degrees in their temperature every day. Whenever this is the case chilly feelings are always produced in those who are exposed to the change, even though the lower temperature curve may not go down to anywhere near freezing. If for instance the temperature at the hottest hour of the day, say three o'clock in the afternoon, is ° F. and that of the later evening or middle of the night is ° F., chilly feelings will be produced. Just the same thing is true if the temperature is between ° F. and ° F. shortly after the middle of the day and then goes down to near zero at night. These chilly feelings are uncomfortable, but they produce an excellent reaction in the circulation and set the blood coursing from the heart to the tissues better than any medicine that we have. In the midst of this the lungs have their resistive vitality raised so as to throw off the disease.

This is probably one of the principal reasons why mountain climates have been found so much more helpful for the treatment of tuberculosis than regions of lower elevations. Whenever the elevation is more than fifteen hundred feet there will almost invariably be a variation of thirty degrees between the day and the night temperature. There are of course still greater variations, even sixty or seventy degrees sometimes where the altitudes are very high, but this is often too great for the tuberculous patients to react properly to, in their rundown conditions. Besides, the air is much rarer at the higher elevations, breathing is more difficult, because the lungs have to breathe more rapidly and more deeply in order to secure the amount of oxygen that is needed for bodily necessities from the rarified air. The middle elevations then, between fifteen hundred and twenty-five hundred feet, have been found the best for tuberculosis patients, and they are very pleasant during the summer time, though never without the chilly discomfort of the drop in temperature. During the fall and winter, however, many patients become tired out trying to react to these variations of temperature and want to seek other climates where they will not have to submit to the discomfort and the chilly feelings. If they come down to more comfortable quarters before their tuberculosis has been brought to a standstill by the increase of their resistive vitality, it is very probable that they will lose most of the benefit that they derived from their mountain experience. Here is where the will comes in. Those who have the will to do it and the persistence to stick at it and the character that keeps them in good humor in spite of the discouraging circumstances which almost inevitably develop from time to time, will almost without exception recover from their tuberculosis with comparatively little difficulty, if they have only taken up the treatment before the disease is so far advanced as to be beyond cure.

In the older days consumptives used to be sent to the Riviera and to Algiers and to other places where the climate was comparatively equable, with the idea that if they could only avoid the chilly feelings consequent upon variations of temperature it would be better for them. Many of the disturbing symptoms of tuberculosis are rendered less troublesome in such a climate, but the disease itself is likely to remain quiescent at best or perhaps even to get insidiously worse, as tuberculosis is so prone to do. These milder climates require much less exercise of the will, but that very fact leaves them without the all-important therapeutic quality which the lower altitudes possess.

For many people the outdoor life and the sight of nature in the variations produced in scenery during the course of the days and the seasons are satisfying enough to be helpful in making their cure of tuberculosis easy. They are extremely fortunate if they have this strong factor in their favor. It is very probable that we owe the discovery of the value of the Adirondacks and other such medium altitudes in the treatment of tuberculosis to the fact that Doctor Trudeau liked the outdoors so much and was indeed so charmed with the Adirondack region that when death from tuberculosis seemed inevitable, he preferred the Saranac region as a place to die in, in spite of the hardships and the bitter cold from which at that time there was so little adequate protection, to the comforts of the city. He scarcely hoped for the miracle of cure from a disease which he as a doctor knew had carried off so many people, but if he were to die he felt that he would rather die in the face of nature with his beloved mountains all around him than in the shut-in spaces of the city.

His resolution to go to the Adirondacks seemed to many of those who heard of it scarcely more than the caprice of a man whom death had marked for itself. His physicians surely had no hope of his journey benefiting him but they felt very probably that in the conditions he might be allowed to have this last desire since there were so few other desires of life that he was likely to have fulfilled. His will to live outdoors in spite of the bitter cold of that first winter undoubtedly saved his life and then he evolved the system of outdoor treatment which has in the past fifty years saved so many lives and is now the recognized treatment for the disease. It is easy to understand, however, how much of firm determination was required on his part forty years ago, when there were no comfortable ways of getting into the Adirondacks, when the last stage of the journey had to be made for forty miles on a mattress in a rough wagon, when water for washing had to be secured by breaking the ice in the pitcher or on the lake and when the bitter climate must have been the source of almost poignant torture to a man constantly running a slight temperature. He had the courage and the will power to do it and the result was not only his own survival but a great benefit secured for others.

Unfortunately many a consumptive patient who during his first period of treatment keeps to the letter the regulations for outdoor air and abundant food fails to do so if it is necessary to come back a second time. Persistency is here a jewel indeed and only the persistent win out. Many an arrested case fails to keep the rules of living that may be necessary for years afterwards and runs upon relapse. The will to do what is necessary is all-important. Trudeau himself, after securing the arrest of his disease in the Adirondacks, though he lived and worked successfully to almost seventy years of age, found it quite impossible to live out of them and often had to hurry back from even comparatively brief visits to the lowlands. Besides, every now and then during some forty years he had the will power to take his own prescription of outdoor air and absolute rest. It was the faculty to do this that gave him length of life far beyond the average of humanity and the power to accomplish so much in spite of the invasion of the disease which had rendered large parts of both lungs inoperative. Not only did he live on, however, but he succeeded in doing so much valuable work that few men in the medical profession of America have stamped their name deeper on modern medical science than this consumptive who had constantly to use his will to keep himself from letting go.

CHAPTER XII

THE WILL IN PNEUMONIA

"Who shall stay you? My will, not all the world."
Hamlet

What is true of tuberculosis and the influence of the will has proved to be still more true, if possible, of pneumonia. Clinical experience with the disease in recent years has not brought to us any remedy that is of special value, nor least of all of specific significance, but it has enabled us to understand how individual must be the treatment of patients suffering from pneumonia. We have recognized above all that mentally disturbing factors which lessen the patient's courage and will to live may prove extremely serious. We hesitate about letting an older person suffering from pneumonia learn any bad news and particularly any announcement of the death of a near relative, above all, a husband or wife. The shock and depression consequent upon any such announcement may prove serious or even fatal. The heart needs all its power to accomplish its difficult task of forcing blood through the limited space left free in the unaffected lung tissue, and anything which lessens that, that is anything which disheartens the patient, to use our expressive English phrase, must be avoided as far as possible.

When a man of fifty or beyond, one or more of whose friends has died of pneumonia about his age, comes down with the disease and learns, as he often will in spite of the best directed effort to the contrary, that he is suffering from the affection, if he is seriously disturbed by the knowledge, we realize that it bodes ill for the course of the disease. If a pneumonia patient, especially beyond middle life, early in the case expresses the thought that perhaps this may be the end and clings at all insistently to that idea, the physician is almost sure to feel little confidence of pulling him through the illness. In probably no disease is it more important that the patient's courage should be kept up and that his will should help rather than hamper.

Courage is above all necessary in pneumonia because the organs that are most affected and have most to do with his recovery are so much under the control of the emotions. Any emotional disturbance will cause the heart to be affected to some extent and the respiration to be altered in some way. When a pneumonia patient has to lie for days watching his respirations at forty to the minute, though probably he has never noticed them before, and feels how his heart is laboring, no wonder that he gets scared, and yet his scare is the very worst thing that can happen to him. It will further disturb both his heart and his respiration and leave him with less energy to overcome the affection. He may be tempted to make conscious efforts to help his lungs in their work, though any such attempt will almost surely do more harm than good. He must just face the inevitable for some five to nine days, hope for the best all the time and keep up his courage so as not to disturb his heart. After middle life only the patients who are capable of doing that will survive the trial that pneumonia gives. The super-abounding energy of the young man will carry him through it much better; and besides, the young man usually has much less solicitude as to the future and much less depending on his recovery.

A generation ago or even less, whiskey or brandy or some form of strong, alcoholic stimulant, as it was called, was looked upon as the sheet anchor in pneumonia. For a generation or more at that time, the same remedy had been looked upon by a great many physicians as an extremely precious resource in the treatment of tuberculosis. The therapeutic theory behind the practice was that in affections of the lungs a particular strain was placed upon the heart and therefore this organ needed to be stimulated just as far as could be done with safety. As alcohol increases the rapidity of the heart beat, it was considered to be surely a stimulant and came to be looked upon as the safest of heart stimulants, because, except when used over very long periods, direct bad effects had not been noticed. In pneumonia, above all, the heart needed to be stimulated because it had to pump blood through the portion of the lungs unaffected by the pneumonia, usually congested and offering special hindrances to the circulation; besides, a much larger amount of blood than usual had to be pumped through these portions of the lungs in order to compensate for the solidified portions.

A number of very experienced physicians came to be quite sure that alcoholic stimulants were the most valuable remedy that we had for this special purpose of cardiac stimulation; some of them went so far as to say, with a well known New York clinician, that if they were to be offered all the drugs of the pharmacopeia without alcoholic stimulant for the treatment of pneumonia on the one hand, or whiskey or brandy on the other without all the pharmacals, they would prefer to take the alcohol, confident that it would save more patients for them. They were quite sure that they had made observations which justified them in this conclusion.

We know at the present time that alcohol is not a stimulant but always a narcotic. It increases the rapidity of the heart beat, though not by direct stimulation, but by disturbing the inhibitory nerve apparatus of the heart and thus permitting the heart to beat faster. Just as there is a governor on a steam engine, to keep it from going too fast and regulate its speed to a definite range, so there is a similar governing apparatus or mechanism in connection with the heart. It is by affecting this that alcohol makes the heart go faster. Blood pressure is not raised, but on the contrary lowered, and the effect of alcohol is depression and not stimulation. In spite of this, good observers seemed to note favorable effects from the use of alcohol in both pneumonia and tuberculosis. This appears to be a paradox until one analyzes the psychic effects of alcohol and places them alongside the physical, in order to determine the ultimate equation of the influence of the substance.

Alcohol has a very definite tendency to produce a state of euphoria, that is, of well-being. The patient's mind is brought to where it dismisses solicitude with regard to himself. This neutralizes directly the anxiety which so often acts as a definite brake upon resistive vitality. The alcoholic stimulant, in so far as it has any physical effect, probably does a little harm, but its influence on the mind of the patient not only serves to neutralize this, but adds distinctly to the patient's prospects of recovery. Without it, the dread which comes over him paralyzes to some extent at least his heart activity and interferes with lung action. Under the influence of alcohol, he gains courage artificial, it is true but still enough to put heart in him, and this is the stimulation that the older clinical observers noted. The patient can, with the scare lifted, use his will to be well ever so much more effectively and psychic factors are neutralized that were hampering his resistive vitality.

This illustrates very well indeed the place of dilute alcohol in some of the usual forms in therapeutics about the middle of the nineteenth century. Practically all the textbooks of medicine at that time recommended alcohol for many of the continued fevers. In sepsis, in child-bed fever, in typhoid, in typhus, as well as in tuberculosis and pneumonia and other less common affections, whiskey or brandy was recommended highly and usually given in considerable quantities. All of these affections are likely to be accompanied by considerable anxiety and solicitude with a series of recurring dreads that sadly interfere with nature's efforts toward recovery. Under certain circumstances, the scare, to use the plain, simple word, was sufficient to turn the scale against the patient. The giving of whiskey at least lifted the scare [Footnote]   and enabled the patient to use his vital resources to best advantage.

[Footnote : The use of whiskey for snake-bite probably has no other significance than this lifting of the scare. It used to be said that the alcoholic stimulation neutralized the depressant effect of snake poisoning on the heart. Now we know that this is not true, and in addition, we know of no effect that alcohol in the system might have in neutralizing the presence of the toxic albumin which constitutes the danger in snake poisoning. It is only rarely that the bite of a rattlesnake will be fatal. Experts declare that the snake must be a large one, its sting must be inflicted on the bare skin, it must not have stung any one so as to empty its poison glands for more than twenty-four hours, and the full dose of the poison must be injected beneath the skin for the bite to be fatal. Very rarely are all these conditions fulfilled. When a person is bitten by a snake, however, the terror which ensues is quite sufficient of itself to hurt the patient seriously and he may scare himself to death, though the snake poison would not have killed him.

The whiskey lifts the scare and gives nature a chance to neutralize the poison which she can usually do successfully.]

It is extremely important, then, first to be sure that the patient's will to be well is not hampered by unfortunate psychic factors and secondly, that his courage shall be stimulated to the greatest possible degree. Fresh air is the most important adjuvant for this that we have. The outdoor air gives a man the courage to dissipate dreads and makes him feel that he can accomplish what seemed impossible before. Undoubtedly this is one of the favorable effects of the fresh-air treatment of pneumonia, for it makes people mentally ever so much less morbid. The patient's surroundings must be made as encouraging as possible and there must be no signs of anxious solicitude, no long faces, no weeping, and as far as possible, no disturbance about business affairs that might make him think that a fatal termination was feared. His will to get well must be fostered in every possible way and obstacles removed. This is why it has been so well said in recent years that good nursing is the most important part of the treatment of pneumonia. This does not mean that a good nurse can replace a physician, but that both must coordinate their efforts to making the patient just as comfortable as possible, so that he will feel assured that everything that should be, is being done for him, and that it is only a question of being somewhat uncomfortable for a few days and he will surely get well.

Sunny rooms, smiling faces, flowers at his bedside, cheerful greetings, all these, by adding to the patient's euphoria, bolster up his will and make him feel that after all, thousands of people have suffered from pneumonia and recovered from it, and there is no reason why he should not, provided that he will not interfere with his own recovery.

CHAPTER XIII

COUGHS AND COLDS

"The power and corrigible authority of this lies in our wills."
Othello

It might seem as though the will had nothing to do with such very material ailments as coughs and colds, and yet the more one knows about them, the clearer it becomes that their symptoms can be lessened, their duration shortened, their tendencies to complications modified, and to some extent at least, they can be almost literally thrown off by the will to be well. The idea of a little more than a generation ago that coughs and colds would be most benefited by confinement to the house and as far as possible to a room of an absolutely equable temperature has gradually given way before the success of the open-air treatment for tuberculosis and the meaning of fresh air in the management of pneumonia cases. Fresh, cold air is always beneficial to the lungs, no matter what the conditions present in them, though it requires no little courage and will power to face the practical application of that conclusion in many cases. When it is bravely faced, however, the results are most satisfactory, and the respiratory condition, if amenable to therapeutics, is relieved or proceeds to get better. Of course it is well understood that any and every patient who has a rise in temperature, that is whose temperature is above ° F. in the later afternoon hours, should be in bed. Under no circumstances must a person with any degree of fever move around. This does not mean, however, that such patients should not be subjected to fresh, cold air. The windows in their room or the ward in which they are treated should be open, and if the condition is at all prolonged, arrangements should be made for wheeling their beds out on the balcony or placing them close to a window. The cold air gives them distinctly chilly feelings and sometimes they complain of this, but they must be asked to stand it. Of course if the cold disturbs their circulation, if the feet and hands get cold and the lips blue, the patients are not capable of properly reacting against the cold and must not be subjected to it. Their subjective feelings of chilliness, however, must not be sufficient to keep them from the ordeal of cold, fresh air; on the contrary, they must be told of the benefit they will receive from it and asked to exert their wills to stand the discomfort with just as little disturbance as possible.

People suffering from coughs, no matter how severe, should get out into the air regularly, if they have no fever, and should go on with their regular occupation unless that occupation is very confining or is necessarily conducted in dusty air. Keeping to the house only prolongs the affection and makes it much more liable to complications than would otherwise be the case. Sufferers from these affections should not go into crowds, should avoid the theaters and crowded cars, partly for the sake of others because they can readily convey their affection to them but also for their own sake, because they are more susceptible to other forms of bacteria than those already implanted in their own systems and they are much more liable to pick up foreign bacteria in crowds than anywhere else. They should be out in the open air, particularly in the sunlight, and this will do more to shorten the course of a cough and cold than anything else.

They need more sleep than before and should be in bed at least ten or eleven hours in the day, though if they should not sleep during all of that time, they need not feel disturbed but may read or knit or do something else that will occupy them while they retain a recumbent position. They should not indulge in long, tiresome walks and in special exertion, but should postpone these until the cough has given definite signs of beginning to remit.

With regard to the cough itself, it must not be forgotten that the action of coughing is for the special purpose of removing material that needs to be cleared from the lungs and the throat and larynx. It should not be indulged in except for that purpose. It requires a special effort, and while the lungs and other respiratory passages are the subject of a cold, these extra efforts should not be demanded of them unless they are absolutely necessary. Almost needless to say, people indulge in a great deal of unnecessary coughing. Some of this is a sort of habit and some of it is due to that tendency to imitate, so common in mankind. Every one has surely heard during religious services, in a pause just after heads have been bowed in prayer or for a benediction, a single cough from a distant part of the church which seemed to be almost the signal for a whole battery of coughs that followed immediately from every portion of the edifice. If some one begins coughing during a sermon or discourse, others will almost inevitably follow. Coughing, like yawning, is very liable to imitation.

The famous rule of an old-time German physician was that no one was justified in coughing or scratching the head unless these activities were productive. Unless you get something as the result of the coughing, it should not be indulged in. There are a great many people who cough much more than necessary and who delay the progress of their betterment in that way. Whenever material is present to be coughed up, coughing is not only proper but almost indispensable. It is the imitative cough, the coughs which indicate overconsciousness of one's affection, the coughs that so often almost unconsciously are meant to catch the sympathy of those around, which must be repressed by the will, and when the patient finds that he really has to cough less than he thinks, he will be quite sure that he is getting better and will actually improve as a consequence of this feeling.

Coughs need an abundance of fluid much more than medicine, and warm fluids are better than cold; the will must be exercised so as to secure the taking of these regularly. At least a quart of warm liquid, milk if one is not already overweight, should be taken between meals during the existence of a cough. Hot milk taken at night will very often secure much better rest with ever so much less coughing than would otherwise be the case. The tendency to take cough remedies which lessen the cough by their narcotic effect always does harm. Coughing is a necessary evil in connection with coughs, and whatever suppression there is should be accomplished by means of the will. Remedies that lessen the coughing also lock up the secretions and disturb the system generally and therefore prolong the affection and do the patient harm. Most of the remedies that are supposed to choke off a cough have the same effect. Quinine and whiskey have been very popular in this regard but always do harm rather than good. Their use is a relic of the time when whiskey was employed for almost every form of continued fever and when quinine was supposed to be good for every febrile affection. We know now that quinine has no effect except upon malarial fevers, and then only by killing the malarial organism, and that whiskey is a narcotic and not a stimulant and does harm rather than good. Those who did not take the familiar Q. and W. have in recent years had the habit of administering to themselves or to their friends various laxative or anodyne or antiphlogistic remedies that are supposed to abort a cough or cold and above all, prevent complications. All of these remedies do harm. Every single one of them, even if it makes the patient a little more comfortable for the time, produces a condition that prevents the system from throwing off the infection which the cold represents as well and as promptly as it otherwise would.

It requires a good deal of will power to keep from taking the many remedies which friends and sometimes relatives insist on offering us whenever a cold is developing, but the thing to do is to summon the will power and bravely refuse them. Medicine knows no remedies that will abort a cold. The use of brisk purgatives, sometimes to an extent which weakens the patient very much the next day, is simply a relic of the time when every patient was treated with antimony or calomel and free purgation was supposed to be almost as much of a cure-all as blood-letting. There is no reason in the world to think that the emptying of food out of the bowels will do any particular good, unless there is some definite indication that the food material present there should be removed because it is producing some deleterious effect.

The longer a physician is in the practice of medicine the less he tries to abort infectious diseases, and coughs and colds are, of course, just infections. They must run their course, and the one thing essential is to put the patient in as good condition as possible so that his resistive vitality will enable him to throw off the infection as quickly as possible. It requires a good deal of exercise of will power on the part of the physician to keep from running after the many will-o'-the-wisps of treatment that are supposed to be so effective in shortening the course of disease, but any physician who looks back at the end of twenty years will know that his patients have reason to be thankful to him just in proportion as he has avoided running after the fads and fancies of current medicine and conservatively tried to treat his patients rather than cure their diseases. The patient is ever so much more important than his disease, no matter what the disease may be.

Above all, for the cure and prevention of coughs and colds people must not be afraid of cold, fresh air. A good many seem to fear that any exposure to cold air while one has a cough may bring about pneumonia or some other serious complication. It must not be forgotten, however, that the pneumonia months in the year occur in the fall and the spring, October and November and March and April producing most deaths from the disease, and not December, January and February. The large city in this country which may be said to have the fewest deaths from pneumonia is Montreal, where the temperature during December and January is often almost continuously below zero for weeks at a time and where there is snow on the ground for three or four months in succession. The highest death rate from pneumonia is to be found in some of our southern cities which have rather mild winters and rather equable temperature, that is, no considerable variation in the daily temperature range. Cold air is bracing and tonic for the lungs and enables them to resist the microbe of pneumonia, and it is now recognized by physicians that personal immunity is a much more important factor in the prevention of the disease than anything else.

Coughs and colds and bronchitis and pneumonia, the respiratory diseases generally, are much less frequent in very cold climates than in variable regions. Arctic explorers are but rarely troubled by them, even though they may be exposed to extremely low temperatures for months. Men subjected to blizzards at thirty and forty degrees below zero may have fingers and toes frozen but do not have respiratory affections. Some years ago, it was noted that one of these Arctic expeditions had spent nearly two years within the Arctic Circle without suffering from bronchial or throat disease and within a month after their return in the spring most of them had had colds. Nansen and his men actually returned from the Arctic regions where they had been in excellent health during two severe winters to be confined to their beds with grippy colds within a week of their restoration to civilization, with its warm comfortable homes and that absolute absence of chill which is connected in so many people's minds with the thought of coughs and colds.

The principal reason why colds are so frequent in the winter time in our cities and that pneumonia has increased so much is mainly because people are afraid of standing a little cold. Office buildings are now heated up to seventy degrees to make the personnel absolutely comfortable even on the coldest days, and as a consequence the air is so dry that it is more arid that is more lacking in water vapor, as the United States Public

Health Service pointed out than Death Valley, Arizona, in summer. People dress too warmly, anticipating wintry days and often getting milder weather and thus making themselves susceptible to chilling because the skin is so warm that the blood is attracted to the surface. Will power to stand cold, even though at a little cost of discomfort, is the best preventive of coughs and colds and their complications and the best remedy for them, once the acute febrile stage has passed.

CHAPTER XIV

NEUROTIC ASTHMA AND THE WILL

"Great minds of partial indulgence
To their benumbed wills."
Troilus and Cressida

In closing a clinical lecture on bronchial asthma at the University of Marburg some years ago, Professor Friedrich Müller, who afterwards became professor at Berlin, said, "Each asthmatic patient is a problem by himself and must be studied as such; meantime, it must not be forgotten what an important rôle suggestion plays in the treatment of the disease." This represents very probably the reason why so many remedies have been recommended for asthma and have proved very successful in the hands of their inventors or discoverers as regards the first certain number of patients who use them, and yet on subsequent investigation have turned out to be of no special therapeutic value and sometimes indeed to have no physical effect of any significance.

Of course this is said with regard to neurotic asthma only, and must not be applied too particularly to other forms of the affection, though there is no doubt at all that the symptoms of even the most severe cases of organic asthma can be very much modified and often very favorably, by suggestive methods.

The principal feature of asthma is a special form of severe difficulty in breathing. It is known now that the beginning of the affection is always as Strumpell said, "an extensive and quite rapid contraction of the smaller and smallest bronchial branches, that is the terminal twigs of the bronchial tubes." It is not so much air hunger, though there is, of course, an element of that because the lungs are not functioning properly, as an inability to empty the lungs of air already there and get more for respiratory purposes. The spasm in asthma has a tendency to hold the lungs too full of air and produce the feeling of their getting ever fuller and fuller. What the old sea captain said in the midst of his attack of asthma, when somebody sympathized with him because he had so much difficulty in getting his breath, was that he had lots of breath and would like to get rid of some of it. He added, "If I ever get all this breath that's in me now out of me, I'll never draw another breath so long as I live, so help me." The respiration spasm is usually at full inspiration and the effort is mainly directed toward expiration and expulsion of air present using the accessory respiratory muscles for that purpose.

The picture of a man suffering from asthma is that of a patient so severely ill as to be very disturbing to one not accustomed to seeing it. It would be almost impossible for any one not used to the attacks to think that in an hour or two at the most the patient would be quite comfortable and if he is accustomed to the attacks, that he will be walking around the next day almost as if nothing had happened. All that the affection consists in is a spasm of the bronchioles and as soon as that lets up, the patient will be himself again. Some material may have accumulated during the time when the spasm was on which will still need to be disposed of, and there will be, of course, tiredness of muscles unaccustomed to be used in that special way, but that will be all.

We are still in the dark as to what causes the spasms but undoubtedly psychic factors play an important etiological rôle. For a good many people, there is a distinct element of dread as the immediate cause of their asthmatic attacks. Some people have it only when they have gone through some disturbing neurotic experience. Occasionally it is the result of physical factors combined with some psychic element. Cat asthma is not very uncommon and occurs as a consequence of some contact by the individual with a specimen of the cat tribe though usually the large cats, the lions and tigers, do not cause it. There is nearly always, in those who are liable to this form of asthma, a special detestation of cats. There is probably some emanation from the animal which produces the asthmatic fever, just as is true also of horses in those cases where horse asthma occurs. In a few of these latter cases, however, it was noted that the horse asthma did not begin until after there had been some terrifying experience in connection with the horse, as a runaway, a collision, or something of that kind.

Any one who sees many asthmatic cases inevitably gets the impression after a time that their very dread of the attacks has not a little to do with predisposing them. Occasionally the dread is associated with some other organic disturbance, either of heart or kidneys, or oftener still, with some solicitude with regard to these organs and the persuasion that there is something serious the matter with them, though there is at most only some functional disturbance. This is particularly true of cases of palpitation of the heart where there has been considerable dread of organic heart disease. In a certain number of these cases, there is some emphysema present, that is, overdistention of the lungs, such as is seen in high-chested people. Owing to the long anterio-posterior diameter of the chest and the fact that as a consequence it is nearly as thick through as it is wide, this form of chest is sometimes spoken of as barrel-chest. Patients who have it are particularly likely to suffer from asthma if they have any dread of heart trouble or if they are of a nervous constitution.

I have known people with the dread of the dark to get an attack of asthma if they were asked to sleep alone after having been accustomed for years to sleep with somebody in the room. I have known even a physician to have attacks of asthma of quite typical character as the result of a dread of being out after dark which had gradually come over him. I have had a physician patient who was very uncomfortable if alone on the streets of New York, even during the day, and whose symptoms at their worst were distinctly dyspneic or asthmatic. He used to have to bring his wife with him whenever he came to see me for he lived out in one of the neighboring towns, because he was so afraid that he might get an asthmatic attack that would overcome him and he would feel helpless without some one to aid him.

In practically all these cases, the treatment of asthma becomes largely that of treating the accompanying dread. Once the acute symptoms of the attack itself manifest themselves, they have to be treated in any way that experience has shown will relieve the patient. The general condition, however, needs very often an awakening of the will to regulate the life, to get out into the air more than before, to avoid disturbing neurotic elements, and worrying conditions of various kinds. Thin people need to be made to gain in weight, using their will for that purpose; stout people who eat too much and take too little exercise need to have their lives regulated in the opposite direction. In the meantime, anything that arouses the patient to believe firmly that his condition will be improved by some remedy or mode of treatment, will help him to make the intervals between attacks longer and the attacks themselves less disturbing. The will undoubtedly plays a distinct role in this matter which patients who have been through a series of asthmatic attacks recognize very clearly.

The many remedies for asthma which have been lauded highly even by physicians, and that have cured or relieved a great many patients and yet after a while have proved to be without much beneficial effect, make it very clear how much the affection depends on the will power to face it and throw it off. Nothing will be curative in asthma unless the patient has confidence in his power and uses his own will energy to help it. He must overcome the element of dread which occurs in connection with all asthmatic attacks, even those due to organic disease of heart or kidneys. No matter how frequent the attacks have been, there is always an element of fright that enters into an affection which interferes with the respiration. This must be overcome by psychic means to help out the physical remedies that are employed. Sometimes the psychic remedies will succeed of themselves where more material means have failed completely.

CHAPTER XV

THE WILL IN INTESTINAL FUNCTION

"Ill will never said well."
Henry V

During the past generation, the appreciation of the relative part played by the stomach and intestines in digestion has completely changed. Our forefathers considered the stomach the all-important organ of digestion and the intestines as scarcely more than a long tube to facilitate absorption and deal properly with waste materials. Their relative values are now exactly reversed in our estimation. The stomach has come to be looked upon as scarcely more than a thin-walled bag meant to hold the food that we take at each meal and then pass it on by degrees to be digested, prepared for absorption and finally absorbed in the intestines. It has comparatively little to do with such alteration of the food as prepares it to be absorbed. Its motor function is much more important than its secretory function and serious stomach troubles are dependent on disturbances of stomach motility. Contractions at the pyloric orifice, that is the passageway from the stomach into the intestines, will cause the retention of food and seriously interfere with health. The dilatation of the stomach for any reason may produce a like result and these are the stomach affections that need special care.

If the stomach will only pass the food on properly, the intestines will do the rest. A number of people have been found in the course of routine stomach examinations who proved to have no secretory function of the stomach and yet suffered no symptoms at all attributable to this fact. The condition is well known and is called achylia gastrica, that is, failure of the stomach to manufacture chyle, the scientific term for food changed by stomach secretions. Our stomachs are only meant, apparently, to provide a reservoir for food that will save us the necessity of eating frequently during the day, as the herbivorous and graminivorous animals have to do, and enable us to store away enough food to provide nutrition for five or six hours. We thus have the leisure to occupy ourselves with other things besides eating and drinking.

This conclusion as to the relative significance of the stomach and digestion is confirmed by the fact that removal of the whole stomach or practically all of it for cancer has in a number of well known cases been followed by gain in weight and general improvement in health. Schlatter's case, the very first one in which nearly the whole stomach was removed, proved a typical instance of this, for the patient proceeded to gain some forty pounds in weight. She had lost this during the course of the growth of a cancer and its interference with stomach motility. It was necessary, however, for her to be fed, rather carefully, well-chosen foods usually in liquid form, and every hour and a half instead of at longer intervals. Her intestines were thus spared from overloading and proceeded to do the work of digestion for which they are so well provided by abundant secretion poured into them from the large glands, the liver and the pancreas, as well as the series of small glands in their own walls all of which were manifestly meant to do extremely important work.

In the increased estimation of the significance of the digestive functions of the intestines which has come in recent years, there has been a tendency, as always in human affairs, for the pendulum to swing too far. Above all, certain phases of intestinal function have come to occupy too much attention and to be the subject of oversolicitude. Whenever this happens, whatever function it concerns is sure to be interfered with. Attention has been concentrated to a great extent on evacuation of the bowels and the consequences have been rather serious. A great many people whose intestinal functions were proceeding quite regularly have had their attention called to the fact that any sluggishness of the intestines may be the source of disturbing symptoms and the beginning of even serious morbid conditions. As a consequence, they pay a great deal of attention to the matter and before long become so solicitous that the elimination of waste materials from the intestines is interfered with. Above all, they may be led to pick and choose their foods so delicately that there is not the necessary waste material left to encourage peristalsis.

The result is that to some extent at least, intestinal function would almost seem to have broken down in our day. Everywhere one sees advertisements of medicines and remedies and treatments of various kinds that will aid in the evacuation of the bowels. Most of them are guaranteed to be perfectly harmless and all of them are pleasant to take, they work while you are doing nothing else and are just engaged in saving mankind not only suffering but complications of various kinds that may lead to serious results. Some years ago, when Matthew Arnold was in this country, he declared in one of his lectures that what the world needed was "leading and light," but a well known American physician who is closely in touch with American life declared not long since that what we needed in America manifestly, if advertisements were any index of the needs of a people, was laxatives and more laxatives. Advertisements cost money; it is said that at least four times as much as the advertising costs must be spent by the public on any object advertised in order to make it pay, so that very probably nearly a billion of dollars a year is spent in this country on laxatives. Only whiskey and tobacco present a higher bill to the American people annually.

Practically all of the laxative medicines do harm if taken over a prolonged period. Over and over again physicians have found that laxative remedies introduced even by scientists, with the assurance that they were quite harmless and had no undesirable after effects proved the source of annoying or even serious symptoms after a while. It is true that when constipation has become habitual, it may be necessary to give laxative medicines for a prolonged period, but this is only another instance of the necessity that is often presented to the physician of choosing between two evils and trying to find the lesser one. Even the heavy oil that has become so popular in recent years has been found on careful investigation and prolonged observation to have certain undesirable effects and it must not be forgotten that it has not been used generally for a sufficiently long time for us to be absolutely sure what its sequelae may be.

This breakdown of intestinal activity is not the fault of nature but of men and women who have been thinking to improve on the natural laws of living. As the result of improvements in diet and refinements in cooking and the preparation of foods, less and less of their roughage is left in our articles of food when sent to the table. It is on this roughage or waste material that intestinal movement or peristalsis depends. If we eat perfectly white bread, cut all the gristle and fatty materials from our meat, carefully eliminating the connective tissue bundles that may occur in it, eat our vegetables mainly in the shape of purees and avoid to a great extent all the coarser varieties, such as parsnips and carrots and beets, we provide very little material for the intestines to carry on and aid them in the elimination of other wastes. If, besides, we always ride and do not walk, and so have none of that precious jolting which occurs every time the heel comes down, and if we have no bending movements in our lives, no wonder that intestinal movement becomes sluggish and we have to supply stimulants and irritants to get it to do its work.

Intestinal evacuation is very largely a matter of will. There are very few people so constituted by nature that they will not have regular movements sufficient to maintain their digestive tracts in excellent health, if they form the right habits. They must, however, make up their minds, that is their wills, to restore coarse materials to their diet. They must eat whole wheat or graham bread, must eat fruit regularly and usually eat the skins of the fruit with it, that is as far as apples, pears, peaches, apricots, plums and the like are concerned. Even as regards oranges, it is probable that the eating of occasional pieces of orange peel is an excellent means of helping intestinal functions and providing waste material. [Footnote]

[Footnote : A curious discovery has been made in recent years that orange skin contains a very precious element essential for bodily health, belonging to the class of substances known as the vitamines and contains more of it than any other food material that we have. The instinct which tempted so many of us as children to eat orange skin, in spite of the fact that we were discouraged from the practice, was founded on something much more than mere childish caprice. Orange skin is after all the basis of marmalade which has been so commonly used by the English people at breakfast and which is at once a tasty and healthful material.]

When baked potatoes are taken, the skin should be eaten, mainly because of the waste material it provides, but also because just underneath the skin and sure to be removed with it if it is taken off, there are certain salts and other substances that are excellent for health and particularly for digestion. Besides, the carbonized material which so often occurs on baked potatoes is of itself a good thing. It represents some of that charcoal which in recent years French physicians particularly have found very valuable as a remedy for certain disturbances of intestinal digestion. The removal of parings from fruit and vegetables and the careful trimming of meat, have taken out of human diet the materials which meant most for intestinal movements for former generations, and they have to be supplied artificially by means of irritant drugs, salts, oils and the like, to the detriment of function.

The other element in the modern situation as regards the failure of intestinal function is the lack of fluids. People who live indoors are not tempted to take so much water as those who work outside and yet in our modern, steam-heated houses they often need more. Our heating systems take much more water from us than the former methods of heating. The result is seen in our furniture that comes apart from dryness and in our books and other things which crack and deteriorate. Something of the same thing happens to human beings unless they supply sufficient fluids. For this it is necessary deliberately to make up the mind, which always means the will, to consume five or six glasses of water between meals and especially to take one on rising in the morning and another on going to bed. This should not be hot and above all not lukewarm water, but fresh cold water which stimulates peristalsis. The creation of a habit is needed in the matter or it will be neglected. I have sometimes given patients some harmless drug, like a lithium salt, that was to be taken three or four times a day in a full glass of water, in order to be sure that they would take the water. They were willing to take the medicine but I could not be assured that without it they would drink the amount of water that I counselled.

Above all, a regular habit of going to the toilet at a definite time every day must be created. Nothing is so important. In little children, even from their very early years, such a habit can be established; it is only necessary to put them on their chairs at certain times in the day and the desired result will follow. Adults are merely children of a larger growth in this matter, and the habit of going regularly is all-important. A little patience is needed, though there should be no forcing, and after a time, a very satisfactory habit can be established in this manner. It seems almost impossible to many people that anything so simple should prove to be remedial for what to them for a time seemed so serious a disturbance of health, but only a comparatively short trial of the method will be sufficient to demonstrate its value. A book or newspaper may be taken with one, or Lord Chesterfield's advice to learn a page of Horace which may afterwards be sent down as an offering to Libitina, the goddess of secret places, may be followed, but the mind must not be diverted too much from the business in hand, and the will must be afforded an opportunity to exert its power.

It is true that the muscular elements of the intestines consist of unstriped muscles and that they are involuntary, and yet experience and observation have shown that the will has a certain indirect influence even over involuntary muscle. The heart, though entirely involuntary in its regular activities, can be deeply influenced by the will and the emotions, as the words encouraging and discouraging, or the equivalent Saxon words heartening and disheartening, make very clear. Undoubtedly the peristaltic functions of the intestines can be encouraged by a favorable attitude of the will towards them.

Above all, it is important that the anxious solicitude which a great many people have and foster sedulously with regard to the effect of even slight disturbances of intestinal functions should be overcome. We have discussed this question in the chapter on dreads and need only say here that the delay of a few hours in the evacuation of the bowels or even the missing entirely of an intestinal movement for a full day occasionally, will usually not disturb the general health to any notable extent, and that the symptoms so often attributed to these slight disturbances of intestinal function are much more due to the solicitude about them than to any physical effect. There are a great many people whose intestinal functions are quite sluggish and whose movements occur only every second day or so, who are in perfectly good health and strength and have no symptoms attributable to any absorption of supposed toxic materials from the intestines. Indeed, in recent years, the idea of intestinal auto-toxemia has lost more and more in popularity for it has come to be recognized that the symptoms attributed to this condition are due in a number of cases to serious organic disease in other parts of the body, and in a great many cases to functional nervous troubles and to the psycho-neuroses, especially the oversolicitude with regard to the intestines. The will is needed then for intestinal function to regulate the diet, to increase the quantity of fluid, to secure regular habits and to eliminate worry and anxiety which interferes with intestinal peristalsis. There are but very few cases that will not yield to this discipline of the will when properly and persistently tried.

CHAPTER XVI

THE WILL AND THE HEART

"For what I will, I will, and there an end."
Two Gentlemen of Verona

The heart is the primum movens, the first tissue of the body that moves of itself in the animal organism, doing so rhythmically and of course continuously before the nervous system develops in the embryo. This spontaneous activity would seem to place it quite beyond the control of the will, as of course it is, so far as the continuance of its essential activity goes, but there is probably no organ that is so much influenced by the emotions and comes indirectly under the influence of the will as the heart. There are a series of expressions in practically all languages which chronicle this fact. We talk about the encouragement and discouragement or in Saxon terms that are exactly equivalent to the French words, heartening and disheartening of the individual. At moments of panic the heart can be felt to be depressed, while at times when resolve is high there is a sense of well-being in connection with the firm action of the heart that flows over into the organism and makes everything seem easy of accomplishment.

There are a number of heart conditions that depend for their existence and continuance on a sense of discouragement, that is oversolicitude with regard to the heart. If something calls attention to that organ, the fact that it is so important for life and health and that anything the matter with it may easily prove serious, will sometimes precipitate a feeling of panic that is reflected in the heart and adds to the symptoms noted. The original disturbing heart sensation may be due to nothing more than some slight distention of the stomach by gas, or by a rather heavy meal, but once the dread of the presence of a heart condition of some kind comes over the individual, all the subjective feelings in the cardiac region are emphasized and the discouragement that results further disturbs both heart and patient.

Palpitation of the heart is scarcely more than a solicitous noting of the fact that the heart is beating. In certain cases, under the stress of emotion, the heart beat-rate may be faster than normal, but in a number of people who complain of palpitation, no rapid heart action is noted. What has happened is that something having called particular attention to the heart, the beating of the organ gets above the threshold of consciousness and then continues to be noted whenever attention is given it. This is of itself quite sufficient to cause a sense of discomfort in the heart region and there may be, owing to the solicitude about the organ, a great deal of complaint.

Just one thing is absolutely necessary in the treatment of these cases, once it is found that there is no organic condition present. The patient's will must be stimulated to divert the attention from the heart and to keep solicitude from disturbing both that organ and the patient himself. It is not always easy to accomplish this, but where the patient has confidence in the diagnosis and the assurance that nothing serious is the matter, a contrary habit that will overcome the worry with regard to the heart can be formed. For it must not be forgotten that in these cases a series of acts of solicitous attention has been performed which has created a habit that can only be overcome by the opposite habit. It is surprising how much discomfort this simple affection, due to a functional disturbance of the heart and overattention to it, may produce and how much it may interfere with the usual occupation. It is a case, however, simply of willing to be better, and nothing else will accomplish the desired result. At times the mistake is made of giving such patients a heart remedy, perhaps digitalis, but this only emphasizes the unfavorable suggestion and besides, by stimulating heart action, sometimes brings it more into the sphere of consciousness than before and actually does harm.

There is a form of this functional disturbance of the heart which reaches a climax of power to disturb and then is sometimes spoken of as spurious angina pectoris. In these cases the patient complains not only of a sense of discomfort but of actual pain over the heart region and this pain is sometimes spoken of as excruciating. Occasionally the pain will be reflected down the left arm which used to be considered the pathognomic sign of true angina pectoris but is not. Sometimes the pain is reflected in the neck on the left side or at times is noted at the angle of the scapula behind. When these symptoms occur in young persons and particularly in young women, there is no reason to think for a moment of their being due to true angina pectoris, which is a spasm of the heart muscle consequent upon the degeneration of the coronary arteries, the blood vessels which feed the heart itself, and occurs almost exclusively in the old, and much more commonly in old men.

The pain of true angina pectoris is often said to be perhaps the worst torture that humanity has to bear. As a rule, however, it is very prostrating and so genuine sufferers from it are not loud in their complaints. Their suffering is more evident in their faces than in their voices. Indeed, it has come to be looked upon as a rule by the English clinicians and heart experts that the more fuss there is made, the less likelihood there is of the affection being true angina pectoris. When there is pain in the heart region then, especially in young or comparatively young women, of which great complaint is made, it is almost surely to be considered spurious angina, even though there may be reflex pain down the arm as well as the impending sense of death which used to be considered distinctive of the genuine angina pectoris.

The treatment of true angina depends to some extent on inspiring the patient with courage, for it is needed to carry him through the very serious condition to which he is subjected. The psychic element is important, though the drug treatment by the nitrites and especially amyl nitrite is often very effective. In spurious angina, the will is the all-important element. There is some irritation of the heart muscle but it is mainly fright that exaggerates the pain and then concentration of attention on it makes it seem very serious. The one thing that is all important is to relieve patients from the solicitude which comes upon them with regard to their hearts and which prevents them from suppressing their feelings and diverting their minds to other things. Sometimes the will is needed to bring about such a change in the habits of the individual as will furnish proper nutrition for the heart. Very often these patients are under weight, not infrequently they have been staying a great deal in the house, and both of these bad habits of living need to be corrected. Good habits of eating and exercise are above all important for the relief of the condition.

For functional heart trouble, gentle exercise in the open air generally must be taken, for it acts as a tonic stimulant to the heart muscle. Almost as a rule, when patients suffer from symptoms from their hearts, they are inclined to consider them a signal that they must rest and above all must not exercise to such an extent as to make the heart go faster. Rest, if indulged in to too great an extent, has a very unfavorable effect upon the heart, for the heart, like all muscles, needs exercise to keep it in good condition. One of the most important developments of heart therapeutics in our generation was the Nauheim treatment. In this, exercise is an important feature. The exercise is graduated and is pushed so as to make a definite call upon the heart's muscular power. Nauheim is situated in a little cup-shaped valley and patients are directed to walk a certain distance on one of the various roads, distances being marked by signposts every quarter of a mile or so. The walk outward, when the patient is fresh, is slightly uphill, and the return home is always downhill, which saves the patient from any undue strain.

The experience at Nauheim was so favorable that many physicians took up the practice of having their heart patients exercise regularly and found that it was decidedly to their benefit. If this is true for organic heart conditions, it is even more valuable for neurotic heart cases, though it often requires a good deal of exercise of will on the part of patients suffering from these affections to control their feelings and take such exercise as is needed. In men, it will often be found that the discomfort in the heart region, particularly in muscular, well-built men who have no organic condition, is due more to lack of exercise than to any other factor. This is particularly true whenever the men have taken considerable vigorous exercise when they were young and then tried to settle down to the inactive habits of a sedentary life. Athletes who have been on the teams at college, self-made men who have been hard manual laborers when they were young, even sons of farmers who take up city life are likely to suffer in this way. Their successful treatment depends more on getting exercise in the open back into their lives than on anything else, and for this a call upon the individual's will power for the establishment of the needed new habits is the essential.

Former athletes who try to settle down to a very inactive life are almost sure to have uncomfortable feelings in their heart region. At times it will be hard to persuade them that they have not some serious affection consequent upon some overstrain at athletics. In a few cases, this will be found to be true, but in the great majority the root of the trouble is that the heart craves exercise. A good many functional heart cases, like the neurotic indigestions, so called are due to the fact that the heart and the stomach are not given enough to do. The renewal of exercise in the daily life and it should be the daily life as a rule and not merely once or twice a week will do more than anything else to relieve these cases and restore the patient's confidence. We saw during the war that a number of young men, officers even more than privates that is, the better educated more than the less educated suffered from shell shock so called. A good many university men may suffer from what might be termed heart shock if they find any reason to be solicitous about their hearts. These neurotic conditions can only be relieved by the will and diversion of attention.

A certain number of people who suffer from missed beats of their hearts become very much perturbed about the condition of that organ. Irregular heart action, and especially what has been called the irregularly irregular heart, may prove to be a serious condition. There are a number of regular irregularities of heart action, however, consisting particularly of the missed beat at shorter or longer intervals, which may have almost no significance at all. I know two physicians, both athletes when they were at college, who have suffered from a missed heartbeat since their early twenties. In one case it has lasted now for thirty-five years and the physician is still vigorous and hearty, capable even of running up an elevated stairway after a train without any inconvenience. Some twenty years ago there was question of his taking out a twenty-year life insurance policy and the insurance company's physician at first hesitated to accept the risk because of the missed beat. An examination made by three physicians at the home office was followed by his acceptance and he has outlived the maturity of the policy in good health and been given a renewal of it, in spite of the fact that his missed beat still persists.

There is often likely to be a good deal of solicitude as to the eventual prognosis in these cases, that is as to what the prospect of prolonged life is. The regularly irregular heart does not seem to make for an unfavorable prognosis. Young patients particularly who have learned that they have a missed heartbeat need to have this fact emphasized. We have the story of an important official of an American university in whom a missed beat was discovered when he was under forty. This was many years ago, and the prognosis of his condition was considered to be rather serious. The patient actually lived, however, for a little more than fifty years after the discovery of his missed beat. It is easy to understand what a favorable effect on a patient solicitous about a missed beat such a story as this will have. It heartens a patient and gives him the will power to throw off his anxieties and to keep from watching his heart and thus further interfering with its activities. There is even a possibility of life to the eighties or, as I have known at least one case, to the nineties, where the irregular heart was first noted under thirty.

But it is well recognized that close concentration of attention on the heart will hamper its action. It has been demonstrated that it is possible by will power to cause the missing of heartbeats and while only those who have practised the phenomenon can demonstrate it, there are a number of well-authenticated examples of it. There is no doubt, however, that anxiety about the heart will quicken or slow the pulse rate. When a patient comes to be examined for suspected heart trouble the pulse rate is almost sure to be higher than normal, even though there may be nothing the matter with the heart; the increase or decrease of the pulse beat is due to the anxiety lest some heart lesion should be discovered. This makes it necessary as a rule not to take too seriously the pulse rate that is discovered on a first consultation and makes it always advisable to wait until the patient has been reassured to some extent before the pulse rate is definitely taken.

It is easy to see, then, what a large place there is for the will in heart therapeutics. Courage is an extremely important element in keeping the heart from being disturbed and maintaining it properly under control. Scares of various kinds with regard to this all-important organ are prone to get hold of people and then to disturb it. Many a heart that is actually interfered with in its activities by drugs of various kinds would respond to the awakening of the will of the patient so as to control solicitudes, anxieties, dreads and the like that are acting as disturbing factors on the heart. When taken in conjunction with the will to eat and to exercise properly so often necessary in these cases, the will becomes the therapeutic agent whose power must never be forgotten, because it can always be an adjuvant even when it is not curative and can produce excellent auxiliary effects for every form of heart treatment that we have.

CHAPTER XVII

THE WILL IN SO-CALLED CHRONIC RHEUMATISM

"I should do it
With much more ease; for my good will is to it."
The Tempest

In popular estimation, rheumatism is one of the commonest of affections. When a physician asks a patient, especially if the patient is over forty years of age, "Have you ever suffered from rheumatism?" the almost invariable response is, "Yes", though but little further inquiry is needed to show that what the patient means is that he has suffered from some painful conditions in the neighborhood of his joints, or that his muscles have been sore or inclined to ache in rainy weather, or that he has undergone some other vague discomforts connected with dampness. Chronic rheumatism is a term that includes a great many of the most varied conditions. True rheumatism, that is, acute articular rheumatism, is now recognized as an infectious disease which runs a definite course, usually with fever, for some ten days to ten weeks, and requires confinement to bed usually for a month or more. Very rarely will any connection be found between this affection, which presents always Galen's four classic symptoms of inflammation, swelling, redness, heat and pain (tumor, rubor, color, et dolor), and the usual conditions which are broadly characterized as rheumatism. Just as soon as patients are asked if their rheumatism included these symptoms there is denial, yet the idea of their having had rheumatism remains.

As a matter of fact, there are a number of sore and painful conditions in connection with muscles and particularly in and around joints that have, without any scientific justification at least, been called chronic rheumatism. Any painful condition that is worse in rainy weather is sure to be so named. As old dislocations, sprains and wrenches of joints, broken bones, as well as muscular conditions of all kinds, including flat foot and other yielding of joints, all produce this effect, it is easy to understand that there is an immense jumble of all sorts of painful conditions included under the term "chronic rheumatism." Some of them, particularly in older people, produce lameness or at least inability to walk distances without showing the disability; a great many of them produce distinct painful conditions during the night following the use of muscles and often disturb patients very much, because they arouse the dread that they are going to be crippled as they grow older.

Indeed, one of the most serious effects of these recurring painful conditions is the dread produced lest they should cause such progressive affections in and around joints as would eventually make the patients bed-ridden. There are a certain number of cases of so-called rheumatoid arthritis which produce very serious changes in joints with inevitable crippling and quite beyond all possibility of repair. These cases are often spoken of as chronic rheumatism and it is the solicitude produced by the dread of them that makes the worst part of the discomfort in many a so-called chronic rheumatic case. If their affection is to be progressive, then the patients foresee a prolonged confinement to bed in the midst of severe pain, hopeless of ultimate cure. It may be said at once that these cases of rheumatoid arthritis have nothing to do with rheumatism, represent a special acute infection, are never a sequela of any of the rheumatic conditions and are fortunately very rare. This assurance of itself is quite sufficient to make ever so much better a great many patients who feel that they suffer from rheumatism.

The painful conditions that are described under the term chronic rheumatism would seem to be quite beyond any power of the will to affect. They are at least supposed to represent very definite changes in the tissues, usually of chronic character and therefore not amenable to any remedies except those of physical influence. Besides, they are so frequent that surely if there were any question of the will being able to control them or bring relief for them, most sufferers would discover this fact for themselves and apply the remedy from within. It is not to be expected that a very great many people would suffer pains and aches that are worse in rainy weather if all that was needed was the exertion of their will power either to throw off the affection or to perform such exercises and activities as would gradually make their conditions better. In general it is felt that painful conditions of this kind cannot be affected by the will and that distinctly material and not psychic therapeutics must be looked to for their relief.

Now it so happens that the best illustration of the power of the will to "cure" people, that is, to relieve them completely of their affections and start them afresh in life with the feeling that they are no longer handicapped by disease, is to be found exactly in the group of cases that have almost from time immemorial been called chronic rheumatisms. We have had more "cures" of various kinds announced for these chemical, electrical, physical, hydriatic, movement therapy and so forth than for almost any other group of diseases. More irregular practitioners of medicine all down the ages have made a reputation by curing these affections than have won renown by treating any other set of ills to which humanity is heir. Like the poor, these ills are still with us, in spite of all the "cures" and probably nowhere is the expression of the old French physician that "the therapeutics of any generation is always absurd to the second succeeding generation" better illustrated than in regard to them. These cases serve to emphasize very clearly, however, the fact that the pains and aches of mankind are largely under the control of the will.

The more one studies these cases of so-called chronic rheumatism the easier it is to understand how they become the signal "cures" which attract attention to the quacks and charlatans who promise much, but do nothing in particular, though they may give medicines or treatment of some kind or another. They only arouse the patient's will to be better and the determination to use his will with confidence, now that the much praised treatment is doing something which will surely make him better. Cases of this kind have constituted a goodly part of the clientele of the great historic impostors who succeeded in making large sums of money out of curing people by methods that in themselves had no curative power. A review of some of the chapters of that very interesting department of human history, the history of quackery, is extremely suggestive in that regard. The only way to get a good idea of the basic significance of these cases is to realize by what they were cured and by whom they were cured.

One of the most interesting illustrations of that phase of human credulity is the story of Greatrakes, the Irish adventurer who had been a soldier in Flanders, and who when his campaigns were over set up to be a healer of mankind. He chose his opportunity during the time while Cromwell, as Lord Protector of Great Britain, had refused to continue the practice of touching the ailing which the Kings of England had pursued for hundreds of years since the Confessor's time. Cromwell did not impugn the efficacy of the Royal Touch but he refused to have anything to do with it himself. Greatrakes found it an opportune moment to announce that for three nights in succession he had been told in a dream by the Holy Spirit that in the absence of the King he was to touch people and cure them.

One might possibly think that with no better credentials than this and no testimony except his own claim in the matter Greatrakes would receive but scant attention. Any one who thinks so, however, does not understand human nature. It was not long before some of the people who had been sufferers for longer or shorter periods went to Greatrakes and allowed him to try his hand at healing them. They argued that at least if it did them no good it could do them no harm, and it was not long before some of them declared that they had been benefited by his ministrations. Very soon then he was able to furnish what seemed to be abundant evidence of Divine Mission in the cures that were worked by his more than magic touch. Above all, people who had been sufferers for prolonged periods, who had gone the rounds of physicians, who had tried all sorts of popular remedies, and some of whom had been declared incurable were healed of their ills after a series of visits to Greatrakes. No wonder then that patients came more and more frequently, until his name went abroad in all the country and in spite of the difficulties of travel people came from long distances just to be treated by him.

All that he did was to ask the patient to expose the affected part and then Greatrakes would stroke it with his hand, assure the patient that a wonderful new vitality would go into them because of his Mission from on High and promise them that they would surely get better, explaining of course that betterment would be progressive and that it would start from this very moment. The stroking was the important part of the cure and so he is known in history as "Greatrakes the Stroker." It may be said in passing that while those who were touched by the English kings in the exercise of the prerogative of the Royal Touch were usually presented with a gold coin which had been particularly coined for that purpose as a memorial, a corresponding gold piece, a sovereign as a rule, in Greatrakes' method of treatment passed from the patient to the healer. It was a case of metallotherapy with extraction of the precious metal from the patient, as is always the case under such circumstances.

Here in America we had a similar experience, though ours had science as the basis of the superstition in the case instead of religion. The interest aroused by Galvani's experience with the twitching of frogs' legs when exposed nerve and muscle were touched by different metals led Doctor Elisha Perkins to invent a pair of tractors which would presumedly apply Galvani's discovery to therapeutics. These were just plain pieces of metal four or five inches long, shaped more or less like a lead pencil and tapering to a blunt point. With these, as Thatcher, one of our earlier historians of medicine, tells us, Perkins succeeded in curing all sorts of ailments, but particularly many different kinds of painful conditions. He was most successful in the treatment of "pains in the head, face, teeth, breast, side, stomach, back, rheumatism and so forth." In a word, he cured the neuralgias and the rheumatic pains and the chronic rheumatisms which are the source of so much trouble and especially complaint for the old, and which so often physicians, in any time of the world's history, have been unable to cure.

For a time his success was supposed to be due to some curious electrical power that he was using. Learned pamphlets were issued to show that animal magnetism or animal electricity or Galvanism was at work. Professors at no less than three universities in America gave attestations in favor of its efficacy. Time has of course shown that there was absolutely no physical influence of any kind at work. The only appeal was to the mind. Elisha Perkins was a Yale man of education and impressive personality, "possessing by nature uncommon endowments both bodily and mental ", and he succeeded in impressing on his patients the idea that they would surely be cured; he thus overcame the dreads, released the will power, gave new hope and a tonic stimulus to appetite, created a desire for exercise, and then the will kept this up and before long the patient was cured.

When animal magnetism, as it was called about the middle of the nineteenth century, was practised without apparatus, one of its most important claims to the consideration of physicians was founded on its power to heal chronic painful affections which had previously resisted all therapeutic efforts. The power of neuro-hypnotism, as it came to be designated, to accomplish this, will be best appreciated from the fact that this state was being used as a mode of anaesthesia for surgical operations. When the news of the use of ether to produce narcosis for surgical purposes at the Massachusetts General

Hospital first came to England, it did not attract so much attention as would otherwise have been the case, because English physicians and surgeons were just then preoccupied with the discussion of neuro-hypnotic anaesthesia, and those who believed in it thought that ether would not be necessary, while those who refused to believe thought the report with regard to ether just another of these curious self-delusions to which physicians seemed to be so liable.

Perkins' declarations of the curative value of his tractors were, after all, only a succeeding phase of what Mesmer had called to the attention of the medical profession and the public in Paris not quite a generation before. Mesmer seated his patients around a tub containing bottles filled with metallic materials out of which wires were conducted and placed in the hands of patients seated in a circle around it. Mesmer called this apparatus a baquet or battery and it was thought to have some wonderful electric properties. A great many people who received the treatment were cured of chronic pains and aches that had sometimes lasted for years. So many prominent people were involved that the Government finally ordered an investigation to be made by French scientists with whom, because he was the Minister from the colonies at the time, our own Benjamin Franklin was associated. They declared that there was not a trace of electricity or any other physical force in Mesmer's apparatus. He was forbidden to continue the treatment and there was a great scandal about the affair, because a large number of people felt that he was doing a great deal of good.

When hypnotism came in vogue again at the end of the nineteenth century, it was a case of chronic rheumatism that gave it its first impetus in scientific circles. Professor Bernheim of Nancy had tried in vain all of his remedies in the treatment of a patient suffering from lumbago. The patient disappeared for a time and when Bernheim next saw him, he was cured. Bernheim had treated him futilely for months and was curious to know how he had been cured. The patient told him that he had been cured by hypnotism as practised by Liebault. This brought Bernheim to investigate Liebault's method of hypnotism and made him a convert to its practice. It was the interest of the school of Nancy in the subject that finally aroused Charcot's attention and gave us the phase of interest in hypnotism which attracted so much public attention some thirty years ago. Many other cases of those very refractory affections lumbago and sciatica have been cured by hypnotism when they have resisted the best directed treatment of other kinds over very long periods.

It is these chronic rheumatisms, so called, the chronic pains and aches in muscles in the neighborhood of joints, that were cured by the Viennese astronomer, Father Maximilian Höll, in the eighteenth century. He simply applied the magnet and saw the result, and felt sure that there must be some physical effect, though there was none. His work was taken up by Pfarrer Gassner of Elwangen who, after using the magnets for a time, found that there was no need of their application, provided the patients could by prayer and other religious means be brought into a state of mind where they were sure that they were going to get better. They then proceeded to use their muscles properly in spite of the pain that might result for a time, and as a result it was not long before they were cured of their affections. The Church forbade his further practise because of his expressed idea that pain came from the power of evil and dropped from men when they turned to God, which was the eighteenth-century anticipation of Eddyism. Dowie's cures were largely of similar affections, and patients sometimes dropped their crutches and walked straight who could not walk before.

A great many of the so-called chronic rheumatisms are really the result of dreads to use muscles in the proper way because for the moment something has happened to make their use painful. A direct injury, a wrench, or some incident causes a joint for a time to be painful when used. In sparing it, the muscles around it are used differently than before and as a consequence become sensitive and painful. It is quite easy, then, for people to form bad habits which they cannot break because they have not the strength of will to endure the sore and tender condition which develops when they try to use muscles properly once more. The young athlete who wants to get his muscles in good condition knows that he must pass through a period of soreness and tenderness, sometimes of almost excruciatingly painful character. He does so, however, and does not speak of his condition as involving pains and aches but only soreness and tenderness.

Older people, however, who have to get their muscles back into good condition after a period of disuse following an injury or some inflammatory disturbance, find this period of discomfort very difficult to bear and so keep on using their muscles somewhat abnormally and at mechanical disadvantage. As a consequence, these muscles remain tender, are likely to ache in rainy weather and often give a good deal of discomfort. Until the sufferers can be brought to use their wills properly, so as to win back their muscles to normal use, they will not get well. An application of magnets or a Leyden jar or Mesmer's battery of the eighteenth century, or Perkins' tractors, or neuro-hypnotism, or animal magnetism, or later hypnotism, or Dowie's declaration of their cure, enables them to use their will in this regard and then they proceed to recover. It is surprising how many presumedly intelligent people at least they have received considerable education have been cured of conditions that they have endured for years by some remedy or mode of treatment that actually had no physical effect.

St. John Long, the English charlatan who has been mentioned in the chapter on tuberculosis, also succeeded in making a name for himself in connection with the chronic rheumatisms and the so-called rheumatic pains and aches of older people. Between consumption and these conditions, he caught both the young and the old, and thus rounded out his clientele. For consumption he provided an inhalant; for rheumatic conditions, a liniment. This liniment became very famous in that generation for its power to relieve the pains and aches, both acute and chronic, of mankind. So many people were cured by it and above all, so many of them were people of distinction lords and ladies and the relatives of the nobility that Parliament was finally petitioned in the interests of suffering humanity to buy the secret of the liniment from its inventor and publish it for the benefit of the world. I believe that a substantial sum, representing many, many thousands of dollars in our time, was actually voted to St. John Long and the recipe for his liniment was published in the British pharmacopeia. In composition, it was, I believe, only a commonplace turpentine liniment made up with yolks of eggs instead of oil, as had been the custom before. Just as soon as this fact became known, the wonderful cures which had occurred in connection with its use ceased to a great extent, for distinguished members of the nobility and their relatives would not be cured by so common-place a medium as an ordinary turpentine liniment. St. John Long was even accused of not having sold his real secret to the Government, but there was no reason at all to think that. He had been producing his cures not by his liniment but by the strong effect of his prestige and reputation as a healer upon the minds of his patients and the consequent release of will power which enabled them to do things which they thought they could not do before. We have had many wonderful curative oils of various kinds since then, with all sorts of names from Alpha to Omega and very often called after a saint, though St. John Long was as far as possible from being a saint in the ordinary acceptance of that word. These modern curative oils and liniments have been merely counter-irritants, but at times, owing to a special reputation acquired, they have been counter-irritants for the mind and stimulants for the will which have enabled old people to persist through the periods of soreness and tiredness until they reacquired the proper use of their muscles.

CHAPTER XVIII

PSYCHO-NEUROSES

"Look, what I will not, that I cannot do."
Measure for Measure

The psycho-neuroses, that is, the various perversions of nervous energy and inability to supply and conduct nervous impulses properly, consequent upon a mental persuasion which interferes with these activities, have come to occupy an ever larger and larger place in the field of medicine. The war has been illuminating in this matter. A psycho-neurosis is, after all, a hysterical manifestation and it might very well be expected that very few of these would be encountered in armies which took only the men of early adult life and from among those, only persons who had been demonstrated to be physically and, as far as could be determined, mentally normal. Neurologists would seem scarcely to have a place in the war except for wounds of nerves and the cerebral location of missiles and lesions. Certainly none of the army medical departments had the slightest premonition that neurology would bulk larger in their war work than any other department except surgery. That proved to be the case, however.

The surprise was to have, from very early in the war, literally thousands of cases of psycho-neuroses, "shell shock" as unfortunately they came to be called, which included hysterical symptoms of all kinds, mutism, deafness, blindness, paralysis, and contractures. France and England after some time actually had to maintain some fifty thousand beds in their war hospitals mainly for functional nervous diseases, the war neuroses of many kinds. During the first half of the war, one seventh of all the discharges from the British army or actually one third of all the discharges, if those from wounds were not included, were for these war neuroses. They attacked particularly the better educated among the men and were four times as prevalent among officers as among the privates. In proportion to the whole number of those exposed to shells and "war's alarms and dangers" generally, these war neuroses were more common among the men than among the women. Nurses occasionally suffered from them, but not so frequently as the men who shared their dangers in the hospitals and stations for wounded not far from the firing line.

In the treatment of this immense number of cases, a very large amount of the most valuable therapeutic experience for psychoneuroses was accumulated. It was found that suggestion played a very large role in making the cases worse. If these patients were placed in general hospitals where there was much talk of wounds and injuries and the severe trials of battle life they grew progressively worse. They talked of their own experiences, constantly enlarging them; they repeated what they had heard from others as if these represented their own war incidents and auto-suggested themselves into ever worse and worse symptomatic conditions. This was, after all, only the familiar pseudologia hysterica which occurs in connection with hysteria, and which is so much better called by the straightforward name of pathological self-deception or perhaps even just frankly hysterical lying. If these patients were examined frequently by physicians, their symptoms became more and more varied and disabling and their psycho-neurosis involved more external symptoms.

In a word, it was found that their minds were the source of extremely unfavorable factors in their cases. The original shock or the severe trials of war life had unbalanced their self-control and suggestions of various kinds made them still worse. Much attention to their condition from themselves and others simply proved to be constantly disturbing. As was pointed out by Doctor Pearce Bailey, who had the opportunity as United States Chief of the Division of Neurology and Psychiatry attached to the Surgeon General's office to visit France and England officially to make observations on the war neuroses, the experience of the war has amply confirmed Babinski's position with regard to hysteria. The distinguished French neurologist has shown that the classic symptoms of hysteria are the results of suggestion originating in medical examinations or from misapplied medical or surgical treatment. He differs entirely from Charcot in the matter and points out that it was unfortunate misdirected attention to hysterical patients which led to the creation of the many cases of grande hystérie which used to be seen so commonly in clinics in France and have now practically disappeared. They were not genuine pathological conditions in any sense of the word, but merely the reflection of the exaggerated interest shown in them by those interested in neurology, who came to see certain symptoms and were, of course, gratified in this regard by the patients, always anxious to be the center of attention and, above all, the focus of special interest.

The successful treatment of the war neuroses was all founded on the will and not on the mind. Once a careful examination had determined absolutely that no organic morbid condition was present, the patient was given to understand that his case was of no special significance but on the contrary was well understood and had nothing exceptional in it. The unfortunate frequent demonstration of these patients at the beginning of the war as subjects of special interest had been the worst possible thing for them. After experience had cleared the way, they were made to feel that just as soon as the attending physician had the time to give them, he would be able to remove their symptoms without delay. This was almost the only appeal to the mind that was made. It represented the suggestive element of the treatment.

The two other elements were reeducation and discipline. Once suggestion had brought the patient to believe firmly that he would be cured, he was made to understand that his cure would be permanent. Then reeducation was instituted to overcome the bad habit of lack of confidence that had been formed, while discipline broke down the psychic resistance of the patient to the idea of recovery. In such symptoms as mutism or deafness, the patient was told that electricity would cure him and that as soon as he felt the current when the electrode was applied, his power of speech or of hearing would be restored, pari passu, with sensation. The same method was used for blindness and other sensory symptoms. Paralyses were favorably affected the same way, though tremors were harder to deal with. A cure in a single treatment was the best method, for the patient readily relapsed unless he was made to feel that he had recovered his powers completely and that it would be his own fault if he permitted his symptoms to recur.

The most interesting phase of the successful treatment of these war neuroses for us was the fact that the ultimate dependence was placed by the French on a system of management which was called torpillage. Torpillage consists in the brusque application of faradic currents strong enough to be extremely painful in hysterical conditions, and the continuance of the procedure to the point at which the deaf hear, the dumb speak, or those who believe themselves incapable of moving certain groups of muscles come to move them freely. The method has proved highly effective and requires but little time and practically no personnel except the medical officer who applies the treatment and the non-commissioned officer who takes the patient at the end of the treatment and continues the exercise of the afflicted parts. One treatment suffices. The apparatus is of the simplest, the only accessory to the electric supply and the electrodes consisting of an overhead trolley which carries the long connecting wires the whole length of the room, thus making it impossible for the patient to get away from the current which is destined to cure him.

In a word, the man who would insist on maintaining a false attitude of mind towards himself, though that attitude of mind was not deliberate, and least of all not malingering, was simply made to give it up. Sufficient pain was inflicted on him so that he was willing to accept instead of his own false opinion the opinion of his physician that he could accomplish certain functions. Torpillage was, in other words, simply "a method of treatment which gave authority to a medical officer to inflict pain on a patient up to the point at which the patient yields up his neurosis." As a rule, the infliction of very little suffering is needed, for once the demonstration is made that he will have to suffer or give in, it does not take him very long to give in. There is no doubt at all that the method is eminently effective, particularly in those cases which were entirely refractory to other modes of treatment.

It would remind us of some old modes of treatment which were in popular use long ago, but which had gone out entirely in our milder generation because we thought their use almost unjustified. It was not an unusual thing three or four generations ago to rouse a young woman out of an hysterical tantrum, once it was perfectly clear from previous experiences that it was really an hysterical tantrum, by dashing a pitcher of cold water over her. Sir Thomas More relates that he saw a number of people suffering from various forms of possession and any neurologist will confess that some hysterics must have a devil who were cured by being roundly whipped. Certain men and women who complained that they were unable to walk or to work and thus became a care for relatives or for the community, were cured by this, as it seemed to later generations, heartless mode of treatment. Now, we have turned to curing the war hysterias by punishment, that is, by the infliction of severe pain, in just the same way. A great many of these patients who suffer from neuroses and psycho-neuroses, and especially from hysterical inhibitions so that they cannot hear or cannot walk or cannot talk, represent inabilities similar to many which are seen in civil life. Patients complain that they cannot do things; their friends say that they will not do them; and the physician sees that the root of the trouble is that they cannot will. Now, however, that war has permitted the use of such remedies, physicians have found that they can, to advantage, force the patients to will and that once the will has been recalled into action, its energy can be maintained.

Of course the compulsory mode of treatment was not represented as a punishment, but on the contrary it was always presented as a form of treatment which was extremely painful but necessary for the condition. Presented as punishment, it would have been resented, and the patient would probably have set about sympathizing with himself and perhaps seek the sympathy of others, and this would prevent the effectiveness of the treatment. It is very evident that as the result of compulsory methods of treatment, and of the recognition of the fact that major hysterical conditions are largely the result of suggestion and must be cured by enabling the patient to secure control over himself again, the outlook for the treatment of the psychoneuroses will be very different as a consequence of the experience that has been gained. Above all, the place of the will will be recognized, and there will no longer be that coddling of patients and that analysis of their minds for long distant psychic insults of various kinds which will explain their condition, that has done so much harm in a great many ways in recent years.

Another feature of the French treatment was that the neurotic patients should be isolated. This isolation was complete. It had been found that association with other patients, the opportunity to tell their troubles and be sympathized with, did them harm invariably and inevitably, so that those whose neurotic symptoms continued were taken absolutely away from all association with others. Not only this, but all other modes of diversion of mind were denied them. They were placed in rooms without reading or writing materials and even without tobacco. This solitary confinement would remind one of the enforced privacy of the old-fashioned rest cure in which the patient was absolutely secluded from all association with relatives or others who might in any way sympathize with them. The soldier patients were kept in this complete isolation until such time as they showed themselves amenable to treatment. This was usually not very long.

As a matter of fact, the isolation rooms had to be used very little but were found necessary and especially effective in the management of relapsed cases. Just as soon as soldier patients learned that such isolating rooms were available, they became much more ready to give up their neuroses, and as a consequence, in most places, the isolating department did not have to be used, and in some places they could even be given over to the lodgment of attendants. It was quite sufficient, however, that they had fulfilled their purpose of changing patients' attitude of mind towards themselves and giving their will control over them.

As Colonel Pearce Bailey, M.C., says, in most of these patients, persuasive measures and contrary suggestion were quite sufficient, but when they failed, disciplinary measures proved effective. How are we going to be able to make such disciplinary measures available in civil life is another question, but at least the war has made clear that neurotic patients who claim that they cannot do something and actually will not do it, must be made to do it, for this will prove the beginning of their cure. It seems probable, as Doctor Bailey adds, that the reason why the treatment of officers was more difficult and it must not be forgotten that in proportion to their numbers, four times as many officers suffered from so-called shell shock as privates was exactly because these modes of discipline, amounting practically to compulsion, were not used with them.

CHAPTER XIX

FEMININE ILLS AND THE WILL

"Oh, undistinguished space of woman's will!"
King Lear

It is probable that the largest field for the employment of the will for the cure of conditions that are a source of serious discomfort or at least of complaint is to be found among the special ills of womankind. The reason for this is that the personal reaction has so much to do with the amount of complaint in these affections. Not infrequently the individual is ever so much more important than the condition from which she is suffering. Women who have regular occupation with plenty to do, especially if they are interested in it and take their duties seriously, who get sufficient exercise and are out of doors several hours each day and whose appetites are as a consequence reasonably good, suffer very little from feminine ills, as a rule. If an infection of some kind attacks them, they will, of course, have the usual reaction to it, and this may involve a good deal of pain and even eventually require operation. Apart from this, however, there is an immense number of feminine ills dependent almost entirely on the exaggerated tendency to react to even minor discomforts which characterizes women who have no occupation in which they are really interested, who have very little to do, almost no exercise, and whose appetite and sleep as a consequence are almost inevitably disturbed.

Above all, it must not be forgotten that whenever women do not get out into the air regularly every day and this means for a time both morning and afternoon they are likely to become extremely sensitive to pains and aches. This is true of all human beings. Those who are much in the open air complain very little of injuries and bodily conditions that would seem extremely painful to those living sedentary lives and who are much indoors. Riding in the open air is better than not being in the open air at all, but it does not compare in its power to desensitize people with active exercise in the open air. In the older days, when women occupied themselves very much indoors with sewing, knitting and other feminine work, and with reading in the evenings, and when it was considered quite undignified for them to take part in sports, neurotic conditions were even more common than they are at the present time, and young women were supposed to faint readily and were quite expected to have attacks of the "vapors" and the "tantrums."

The interest of young women in sports in recent years and the practice of walking has done a great deal to make them ever so much healthier and has had not a little to do with decreasing the number and intensity of the so-called feminine ills, the special "women's diseases" of the patent medicine advertisements. Much remains to be done in this regard, however, and there are still a great many young women who need to be encouraged to take more exercise in the open than they do and thus to live more natural lives. It is particularly, however, the women of middle age, around forty and beyond it, who need to be encouraged to use their wills for the establishment of habits of regular exercise in the open air as well as the creation of interests of one kind or another that will keep them from thinking too much about themselves and dwelling on their discomforts. These are thus exaggerated until often a woman who has only some of the feelings that are almost normally connected with physiological processes persuades herself that she is the victim of a malady or maladies that make her a pitiable object, deserving of the sympathy of her friends.

A great many of the operations that have been performed on women during the past generation have been quite unnecessary, but have been performed because women felt themselves so miserable that they kept insisting that something must be done to relieve them, until finally it was felt that an operation might do them some good. It would surely do them no harm or at least make them no worse, and there was always the possibility that the rest in the hospital, the firm persuasion that the operation was to do them good, the inculcation of proper habits of eating during convalescence might produce such an effect on their minds as would give them a fresh start in life. Undoubtedly a great many women who were distinctly improved after operations owed their improvement much more to the quiet seclusion of their hospital life, their own strong expectancy and the care bestowed upon them under the hospital discipline without exaggerated sympathy which brought about the formation of good habits of life, than to their operation. Many a woman gained weight after an operation simply because her eating was properly directed, and this was the main part of the improvement which took place.

Operations are sometimes needed and when they are the patient will probably not get well without one; but as a distinguished neurologist, Doctor Dercum of Philadelphia, said in a paper read before the American Medical Association last year, the neurologist is constantly finding patients on whom one or several operations have been performed, some of them rather serious abdominal operations, the source of whose complaints is a neurosis and not any morbid condition of the female or other organs. Occasionally one sees something like this in men, and I shall never forget seeing at Professor Koenig's clinic in Berlin a sufferer from an abdominal neurotic condition on whom no less than three operations for the removal of his appendix had been performed, until finally Professor Koenig felt that he would be justified in tattooing over the right iliac region the words "No Appendix Here." The condition developed in a young soldier as the result of a fall from a horse and his affection resembled very much some of the neuroses that came to be called, unfortunately, "shell shock" during the present war.

The principal trouble in securing such occupation of mind as will prevent exaggerated neurotic reactions to even slight discomforts in women is the creation for them of definite interests in life. The war taught a notable lesson in this regard. Many a physician saw patients whose complaints had been a great source of annoyance to them and their friends proceed to get ever so much better as the result of war interests. In one women's prison in an Eastern State, just before the war, a series of crises of major hysteria was proving almost unmanageable. By psychic contagion it had spread among the prisoners until scarcely a day passed without some prisoner "throwing a fit" with screaming and tearing of clothes and breaking of articles that might be near. Prominent neurologists had been consulted and could suggest nothing. When the war began, the prisoners were set to rolling bandages, knitting socks and sweaters and making United States flags for the army. As if by magic, the neurotic crises disappeared. For months there were none of them. The prisoners had an abiding interest that occupied them deeply in other things besides themselves.

The reduction of nervous complaints of various kinds among better-to-do women was very striking. As might be expected, their rather strenuous occupation with war activities kept them from thinking about themselves, though it is true that now they complain about all the details that they had to care for and the lack of coöperation on the part of certain people. It would seem as though many of them had so much to do that they would surely exhaust their energies and so be in worse condition than before, but this very seldom proved to be the case. Literally many thousands of women improved in health because they became interested in other people's troubles instead of their own. David Harum once said that "It is a mighty good thing for a dog to have fleas because it keeps him from thinking too much about the fact that he is a dog." That seems a rather unsympathetic way of putting the case, but there is no doubt at all that what many women need is serious interests apart from themselves in order to prevent the law of avalanche from making minor ills appear serious troubles.

What most women need above all are heart interests rather than intellectual occupations. That was why occupation with war activities did so much good. That is the reason, too, that club life and reading and other similar pursuits often fail to be helpful to women in their ills to the extent that might possibly be expected. Above all, women need interests in children and the ailing, and these can be supplied by visits to hospitals or by taking an active interest in nurseries, though this is often not personal enough in its appeal to catch a woman's deepest attention. One of the great reasons why there are more nervous diseases among women in our time than in the past is because children are fewer, and because so many women are without children and the calls that they inevitably make on their mothers. Unfortunately, the traditions of the present day are to a great extent in opposition to that family life with a number of children, which means not only the deepest interests for woman but also such inevitable occupations in the care of them that she has very little time to think about herself. It may seem quixotic, that is, demanding unnecessary magnanimity to suggest that these modern ideas should be discarded by those who wish to assure themselves such interests in middle life as will prove definitely preventive of many neurotic conditions, but it is manifestly the physician's duty to make such suggestions.

Life has really become full of dreads for many women in this regard. A gradual reduction in the birth rate which has deprived so many women of the heart interests that were particularly valuable at and after middle life; has been the source of a great deal more suffering without any satisfaction, than would be associated in any way with the care of children. It is extremely unfortunate, then, that this phase of social evolution should have taken place, for the quest of ease and pleasure has proved a prolific source of feminine ills. It is well recognized now that the reason for this reduction in the birth rate is not physical but ethical. It is a matter of choice and not necessity. There is a conscious limitation of the number of children in the family accomplished deliberately, and as a rule the women consider that they are justified in the procedure because they thus conserve their own health and provide such few children as they have with healthier bodies than would otherwise have been the case.

Indeed, child-bearing beyond one or two or perhaps three children has become a source of dread in modern times, a dread that supposedly centers around the health of the children, as well as the mother herself. The mother of a few children is supposed to be healthier and the children of small families to be heartier and more vigorous than when there are half a dozen or more children in the family. A woman is actually supposed by many to seriously imperil her life and her health if she has more than two or three children, though as a matter of fact, the history of the older times when families were larger shows us that women were then healthier on the average than they are now, in spite of all the progress that medicine and surgery have since made in relieving serious ills. Above all, it was often the mother of numerous children who lived long and in good health to be a blessing to those around her, and not the old maids nor the childless wives, for longevity is not a special trait of these latter classes of women. The modern dread of deterioration of vitality as the result of frequent child-bearing is quite without foundation in the realities of human experience.

Some rather carefully made statistics demonstrate that the old tradition in the matter is not merely an impression but a veritable truth as to human nature's reaction to a great natural call. While the mothers of large families born in the slums with all the handicaps of poverty as well as hard work against them, die on the average much younger than the generality of women in the population, careful study of the admirable vital statistics of New South Wales show that the mothers who lived longest were those who under reasonably good conditions bore from five to seven children. Here in America, a study of more favored families shows that the healthiest children come from the large families, and it is in the small families particularly that the delicate, neurotic and generally weakly children are found. Alexander Graham Bell, in his investigation of the Hyde family here in America, discovered that it was in the families of ten or more children that the greatest longevity occurred. So far from mothers being exhausted by the number of children that were born, and thus endowing their children with less vitality than if they had fewer children, it was to the numerous offspring that the highest vitality and physical fitness were given. One special consequence of these is longevity.

In a word, the dread so commonly fostered that the mothers of large families will weaken themselves in the process of child-bearing and unfortunately pass on to their offspring weakling natures by the very fact that they have to repeat the process of giving life and nourishment to them at comparatively short intervals, is as groundless as other dreads, for exactly the opposite is true. It is when nature is called upon to exert her amplest power that she responds most bountifully and dowers both children and mother with better health in return.

Something of the same thing is true with regard to the age of mothers when their children are born. The infant mortality is lowest among the children of young mothers between twenty and twenty-five years of age, though it has been found out that "delay in child-bearing after that age penalizes the children." This is, of course, true particularly for first children. The successive children of young mothers are known by observation and statistics as being constantly in better condition up to the seventh. There is on the average nearly a half a pound difference in weight at birth between succeeding children of the same mother, so that each infant is born sturdier and more vigorous than its predecessor.

These recently collated facts remove entirely the supposed foundations of a series of dreads which were having an unfortunate effect upon our population, for the natives were disappearing before the foreigners because of the higher birth rate among the latter. Birth control has been producing a set of unfortunate conditions for both mothers and children. The one child in the family is sure to be spoiled, not only as a social being but often as regards health, and conditions are scarcely better when there are but two, especially if they are of opposite sexes. If anything happens to them, the mother has nothing to live for, and a little later in life the selfish beings that have been raised under the self-centered conditions of a small family are almost sure to be a source of anxiety and worry. Many a woman owes the valetudinarianism of her later years to the fact that she dreaded maternal obligations and avoided them, and so the latter part of her life is empty of most of what makes life worth living.

The will to make life useful for others rather than to follow a selfish, comfortable, easy existence is the secret of health and happiness for a great many women who are almost invalids or at least constantly complaining in the midst of idle lives. A woman who has nothing better to occupy her time than the care of a dog or two cannot expect to have any interests deep enough to divert her attention from the pains and aches of life that are more or less inevitable. The opportunity to dwell on them will heighten their intensity until they are almost torments. Many more of the feminine ills can be explained in this way than by learned pathological disquisitions. Every physician has seen the bitterest complaints disappear before some change of life that necessitated occupation and gave the patient other things to think about besides self.

The will to face nature's obligations of maternity straightforwardly is probably the greatest preventive against the psycho-neuroses that prove so seriously disturbing to a great many women. Their affections, given a proper opportunity to develop, impel their wills to such activity as prevents the development of morbid states. The dreads for themselves and their children, which so often make the excuse for a different policy in life than this, have proved unfounded on more careful study. Now that war activities no longer call women, it must not be forgotten that home duties are the only ones that can serve as a universal antidote for the poison of self-indulgence, which is much more productive of symptoms of disease than the autointoxications of which we have heard so much, but for which there is so little justification in our advancing science. The assumption of serious duties is the best possible panacea for the ills of mankind as well as womankind, only unfortunately in recent years women have succeeded in shirking duties more and have paid the inevitable price which nature always demands under such circumstances, when the dissatisfaction in life is much harder to bear than the work and trials involved in the pursuit of duty.

Printed in the USA
CPSIA information can be obtained
at www.ICGtesting.com
LVHW080317300524
781288LV00010B/325